Cooperating Expert Systems
in Mechanical Design

ADVANCED SOFTWARE DEVELOPMENT SERIES

Series Editor: **Dr. Jeff Kramer**
Department of Computing
Imperial College of Science, Technology and Medicine
University of London, England

Titles available:

1. Temporal Logic for Real-Time Systems
   Jonathan S. Ostroff

2. Cooperating Expert Systems in Mechanical Design
   Guo Q. Huang *and* **John A. Brandon**

COOPERATING EXPERT SYSTEMS IN MECHANICAL DESIGN

Guo Q. Huang
Engineering Design Centre
School of Engineering
Coventry University
England

John A. Brandon
School of Engineering
University of Wales, Cardiff
Wales

RESEARCH STUDIES PRESS LTD.
Taunton, Somerset, England

JOHN WILEY & SONS INC.
New York · Chichester · Toronto · Brisbane · Singapore

RESEARCH STUDIES PRESS LTD.
24 Belvedere Road, Taunton, Somerset, England TA1 1HD

Marketing and Distribution:

Australia and New Zealand:
JACARANDA WILEY LTD.
GPO Box 859, Brisbane, Queensland 4001, Australia

Canada:
JOHN WILEY & SONS CANADA LIMITED
22 Worcester Road, Rexdale, Ontario, Canada

Europe, Africa, Middle East and Japan:
JOHN WILEY & SONS LIMITED
Baffins Lane, Chichester, West Sussex, England

North and South America:
JOHN WILEY & SONS INC.
605 Third Avenue, New York, NY 10158, USA

South East Asia:
JOHN WILEY & SONS (SEA) PTE LTD
37 Jalan Pemimpin #05-04
Block B Union Industrial Building, Singapore 2057

Library of Congress Cataloging in Publication Data

Huang, Guo Q. (Guo Quan), 1962-
    Cooperating expert systems in mechanical design / Guo Q. Huang,
John A. Brandon.
        p. cm. – (Advanced software development series)
    Includes bibliographical references and index.
    ISBN 0-86380-151-X (Research Studies Press). – ISBN 0-471-94157-3
Wiley
    1. Expert systems (Computer science)   2. Computer-aided design.
3. Design, Industrial.   I. Brandon, John A., 1950- .   II. Title.
III. Series.
QA76.76.E95H83   1993
670'.285'633–dc20   93-24767
                                                                          CIP

British Library Cataloguing in Publication Data

A catalogue record for this book is
available from the British Library.

    ISBN 0 86380 151 X (Research Studies Press Ltd.)
    ISBN 0 471 94157 3 (John Wiley & Sons Inc.)

Printed in Great Britain by SRP Ltd., Exeter

Contents

Preface

This volume attempts to bridge two areas of expertise, those of Mechanical Design/Manufacturing and of Cooperating Expert Systems, both of which are the subject of intensive research. In neither of the two domains are the problems under discussion trivial – indeed they are often so complex that their very definition is a challenging task. As will be seen, this process of problem definition is often the most complex portion of the solution process.

In general the underlying philosophical approaches are radically different. Manufacturing Systems Design is - and is always likely to be - an interdisciplinary activity, and hence is historically based on an extremely eclectic range of ideas, techniques and levels of expertise. In contrast, although the origins of knowledge-based systems are diverse (see, for example, *McCorduck(1979)*), research into new methodologies for expert systems is increasingly becoming the specialist preserve of the computer scientist.

The development of any new specialist discipline passes through a number of stages. Perhaps the most significant of these is the process of the development of specific vocabulary, terminology and notation. Prior to this stage, concepts are expressed in the vernacular and are thus accessible to a broad range of potential contributors and, perhaps more importantly, beneficiaries.

As the discipline matures, knowledge is likely to become less and less accessible, because of its presentation in specialised terminology and notation, at just the time when it starts to become useful to the community at large. Potential users then become dependent on the perception of the specialist that there is merit - either personal (self-interest) or general (altruism) - in the dissemination of knowledge.

Peter Drucker(1989), one of the most influential writers on management, expressed the problem in the following terms:

> '... specialisation is becoming an obstacle to the acquisition of knowledge and an even greater barrier to making it effective. Academia defines knowledge as what gets printed. But surely this is not knowledge, it is raw data. Knowledge is information that changes something or somebody - either by becoming grounds for action, or by making an individual (or an institution) capable of different and more effective action.'

The representation and manipulation of knowledge in abstract forms is necessary for any discipline to progress. For example, the operation on any set of logical arguments of any realistic degree of complexity requires a formal predicate calculus. Thus there appears to be an essential compromise between availability of knowledge in the general community and its usefulness among specialists.

The cultural bias favouring abstraction over applicability, implicit in the criticism of Drucker(ibid), is not a recent development but rather follows a tradition dating back over two thousand years. The ancient Greek geometers placed considerable emphasis on the purity of logical ideas, strongly discouraging any consideration of their applicability. Pure 'Arithmetic' was regarded highly whereas the applied 'Logistic' was disdained. For example, it had long been thought that the geometrical results could not have been developed by the same abstract, and arguably sterile, process which was subsequently used for their presentation. It was not until the early years of the present century, however, that a long-lost manuscript by Archimedes was discovered in Istanbul by the Danish scholar J L Heiberg. This text, 'The Method' illustrates the actual process of conjecture and inference, used to derive the theorems prior to their expression in Euclidean formal proof language:

> 'His other treatises are gems of logical precision, with

little hint of the preliminary analysis that may have led to the definitive formulations. So thoroughly without motivation did the proofs appear to some writers of the seventeenth century that they suspected Archimedes of having concealed his method of approach in order that his work might be admired the more.' *(Boyer 1968)*

At long last, the audience found out how the magician's rabbit got into the hat!

The philosophical approach of the current volume owes much more to the tradition of 'The Method' than it does to other Archimedean texts such as 'On The Sphere and Cylinder' or 'On Floating Bodies' or Euclid's 'Elements'. Emphasis is placed on examining various aspects of the design process rather than on presenting specific design solutions, however elegant. That is not to say that solutions are not regarded as important; indeed a usable prototype Cooperating Expert System will be described at some length.

The authors recognise that purists in either community will be able to point to deficiencies in the text, either by the incorporation of certain simplifying assumptions, which may limit the generality of applicability, or by failure to include the latest technical developments. Such criticism is inevitable where researchers try to make a bridge between two active areas of research. Wherever such devices are utilised reference will be made in the text.

The structure of the volume reflects this approach, making it slightly different from standard textbooks or research monographs. Early chapters (1-3) concentrate on the nature of design prior to the consideration of methodologies for construction of Cooperating Knowledge-Based Systems (chapters 4-6). Chapters 7-10 describe how the theoretical ideas can be put into practice.

It is intended that the book should be accessible to both manufacturing systems engineers and to developers of knowledge-based systems. As a consequence, the pace of the work is relatively gentle. In places concepts will be described both in words and symbolically where other texts would opt for the more concise representation.

Chapter 1
Introduction

1.1 GENERAL MOTIVATIONS

1.1.1 Concurrency in Design and Manufacturing

There is an increasingly diverse demand for a variety of products from manufacturing organisations. This imposes both a requirement for rapid development of new products or improvements in existing products and corresponding advances in the organisation of manufacturing systems and processes. These activities can be complex, time-consuming, and expensive. Resulting decisions may have lasting impacts on the viability and competitiveness of the enterprise, and commonly involve conflicting economic, technical, and social constraints. To make well-informed decisions, developers must obtain sufficient and suitable information from disparate sources, and solicit advice from various functional specialists, both within and outside the company. This data and advice may be incorporated into a variety of models in performing comparative analysis of different candidate development plans.

The term manufacturing design is used in this volume in its broadest sense, to emphasise the integration of product design and manufacturing processes into manufacturing systems. As a consequence it encompasses the following three general aspects:

- design of products
- design of manufacturing systems
- design of manufacturing processes

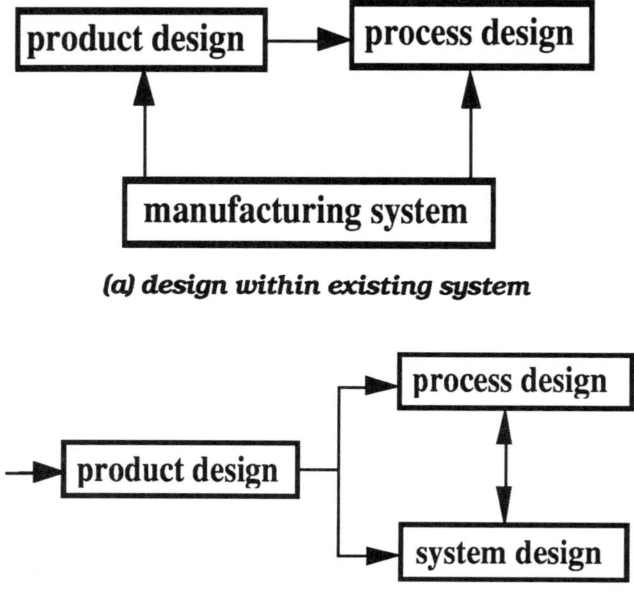

(a) design within existing system

(b) design within new system

Fig. 1.1 traditional design of products, processes, and systems

The central theme of manufacturing design has recently been subsumed within a more comprehensive framework of concurrent or simultaneous engineering. Concurrent engineering is widely regarded as a new approach to product development although it can be argued convincingly that it is the default organisational system in the entrepreneurial company *(Brandon and Huang 1993)*. By applying theories and techniques from systems engineering in design problem solving, it incorporates numerous factors related to constituent components of an artifact throughout its life cycle. The ultimate objective of concurrent engineering is, therefore, to solve realistically large problems, ensuring total quality and reducing project duration whilst minimising cost. At the topmost level, concurrent engineering integrates the three general aspects: system design, product design and process design, into one coordinated activity. Traditionally, these aspects have been arranged sequentially. This is indicated in figure 1.1, with two variations. The first concerns the design of a product within an existing manufacturing system. The second more general problem requires both the design of a product and consequent synthesis of a suitable manufacturing system. In both cases, however, process design implies the synthesis of the mechanism whereby raw materials are converted to a product meeting customer requirements.

Within concurrent engineering, although much of the sequential characteristic is necessarily retained, feedback from downstream activities is used at as early a stage as possible. This is indicated in figure 1.2. To summarise, there are three basic types of problems which concurrent engineering can deal with effectively (a real problem may be a combination of these):

(1) The artifact is fairly primitive, consisting of only one single component of itself. However, the problem must be dealt with from a number of points of view.
(2) The artifact consists of a number of primitive components. However, design of individual components influences the solution space of the original problem.
(3) The artifact passes through a series of inter-relating phases during its life span.

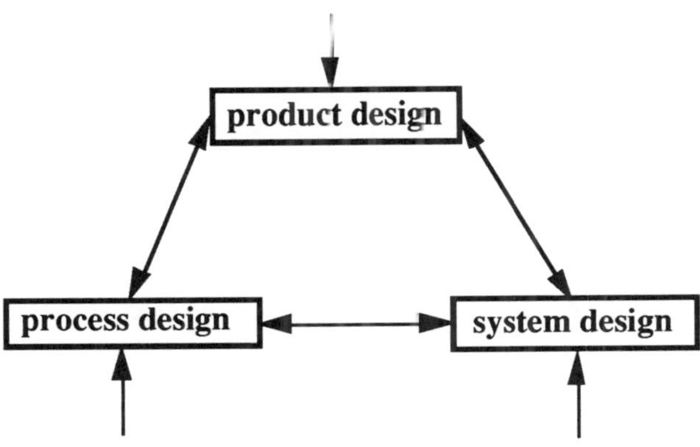

Fig. 1.2 concurrent design of products, processes, and systems

1.1.2 From Computer-Aided Design to Expert Design Systems

Computers are being exploited increasingly to cope with the complexity of system/product/process development. Two of the most successful applications of Computer Aided Design (CAD) are Computer Aided Geometric Design (CAD-G) and Computer Aided Analytical Design (CAD-A). They have played an important role in product design, especially during the later stages such as detailing. However, these two approaches have their limitations. For instance, there is a need for strict mathematical models and accurate data in analytical design, which may not be available during early phases such as conceptual configuration. Pure geometric data are inadequate to support other activities such as process planning and design for manufacture *(Shaw et al 1989)*.

Expert Systems (ES) *(Hayes-Roth et al 1983)*, based on ideas from Artificial Intelligence (AI) *(Nilsson 1980)*, are expected to extend the capability of existing CAD systems. Two approaches have been attempted for developing Intelligent CAD (CAD-I) systems *(Smithers 1989)*. The methodology of the short-term approach is to extend existing CAD systems using AI techniques. This kind of embellishment has been criticised as being insufficient. The long-term approach is to develop AI-based design support systems. In this latter approach, also adopted in this volume, AI techniques are used not just to improve performance but also to extend the capability.

1.2 HUMAN EXPERTS AND EXPERT SYSTEMS

1.2.1 Dichotomy: "Narrow Domains" and "Complicated Problems"

The intuitive expectation of an expert system is that it will perform its tasks in a manner similar to that of a human expert solving a problem in the domain of application. Other expectations include, for example, solving problems that cannot be tackled by the analytical or graphical approach. A large number of systems for research and/or applications have appeared, though most of them are prototype versions. *Heragu and Kusiak (1987)* have reviewed some of the early systems, in terms of how they are designed, what they are composed of, and how they work. How successful these systems are for real applications, unfortunately, has yet to be demonstrated. This is not within the scope of this research. However, what is to be emphasised here is the following dichotomy:

AI scientists	design engineers
Expert systems are applicable to the solution of problems within narrow domains.	Expert systems have been used to solve problems which are very complicated.

Careful explanation must therefore be given of the meanings of "narrow" and "complicated". A complicated problem may fall into a narrow domain, whereas a problem with a wide domain may not be necessarily complicated. In the present work such issues will not be addressed, as it is contended that such semantic nuances can be avoided by defining such terms contextually. However, researchers must be aware of the limited extent to which expert systems have been used to solve real engineering problems. The Cooperating Expert Systems (CES) approach discussed in this research has the potential to provide sufficiently sophisticated analytical capability to address prevailing problems in concurrent engineering

1.2.2 Primitive Activities Constitute Complexity

Most of the Knowledge-Based (KB) expert systems in mechanical design and manufacturing are for essentially primitive activities. By this it is meant that:

(1) They deal with simple products such as gear design and bearing selection.
(2) They are concerned with single aspect of the problem, often from a single viewpoint, for example structural performance or geometric properties.
(3) They consider a problem at a particular stage such as detailing design and production.
(4) They employ simple techniques based on straightforward theory.

It is these primitive activities that constitute a complicated problem in reality. The central problem is whether it is possible to integrate these systems to solve a larger problem whose domain envelops the individual systems. For example, with KB systems for gear design, shaft design, bearing selection, lubricant selection, etc., a system for designing a transmission box is of special concern.

1.2.3 Teamwork versus Individual Contributions

Research over the last decade has revealed that design is based on teamwork, as pointed out by *Pahl and Beitz (1977, page 3)*:

> "In both psychological and organisational respects, design work on a particular problem and in a particular place is performed by people with extremely varied education and experience. For the most effective division of labour, the work should be shared out between academically trained engineers, engineers with a more practical background, technicians and draftsmen, all working in close collaboration."

Most activities in manufacturing design are not usually carried out by a single expert alone. A project is actually led by a single expert, or perhaps a generalist. A number of specialists are involved in a team or group established for the project. They complete a project with mutual assistance and constraints. This indicates that cooperation is an essential ingredient of manufacturing design.

Differentiated human design teams with specialised members should ideally be organised in such a way that an optimum degree of cooperation can be achieved. Suppose that individual designers in a team are assisted by computers, it is of practical concern how these computer-aided workstations should be organised to coordinate the solution of a common problem.

1.2.4 Distributed Design and Centralised Expert Design Systems

Manufacturing design is characterised by its distributed nature. Cooperation between problem solvers is, therefore, a natural consequence of distributed complexity. Distribution and cooperation in manufacturing design can be interpreted in a number of ways, as summarised as follows:

- Design is distributed in the sense that a product usually consists of a number of components distributed as a system to provide desired functions.
- Design is distributed in the sense that a product must be considered from multiple viewpoints to ensure the total quality.
- Design is distributed in the sense that human design organisations are differentiated, and team members are specialised to extend the collective capacity and capability of human cognition.
- Design is distributed in the sense that a product evolves from qualitative and incomplete specifications to quantitative and exact descriptions through a number of stages with changing emphasis.
- Design is distributed in the geographical sense. However, this point may be considered as trivial since geographically separated design experts can be gathered together at some price such as travel, accommodation, and so on.

Most early expert systems developed for mechanical design and manufacturing imitate architectures of early diagnostic systems typified by MYCIN *(Shortliffe 1976)* and PROSPECTOR *(Duda, Gaschnig and Hart 1979)*. They are commonly simple rule-based systems, which are not adequate to represent variety of knowledge and information, especially in a domain such as manufacturing design *(Brown 1984)*. They have, in general, failed to incorporate the cooperative nature of problem solving in manufacturing design, though some systems tend to, or are forced to, organise their knowledge in such a fashion. Consequently, resulting systems cannot be expected to solve "real" and "large" problems in such a way, and to such an extent, that a human expert or a group of experts does.

1.3 DISTRIBUTED AI AND COOPERATING EXPERT SYSTEMS

Third-Party Development: Due to limitations in both human and machine capabilities of cognition and memory, it is not surprising that systems for solving sub-problems constituting an original global problem are commonly developed locally and in isolation by different groups of human experts. In contrast to the top-down approach, where the global objective is defined first, leading to a structured and logical development of problem specification, different groups of people may develop their systems through a bottom-up approach without necessarily being aware of how useful their encapsulated knowledge and expertise are to others. In either case, system designers should keep in mind the possibility that their systems may be useful to others despite the fact that no immediate linkages may be apparent.

Use of Proprietary Expertise: A number of individual systems for primitive design activities have been reported which work well within specified narrow domains. It is worthwhile to utilise the existing expertise accumulated among these successful systems wherever this is practicable. On the other hand, it is also possible, and relatively common, to develop a traditional (rule-based) expert system especially for a design problem, regardless of equivalent expertise already captured within existing expert systems for component design. However, this latter approach is in conflict with the principle of third-party development.

Representing Common Knowledge: To enable cooperation, there should be a protocol which provides a general consensus for the representation of common knowledge. Although expertise and knowledge captured within individual systems may well share overlapping ingredients, this may not be explicitly represented. Even worse, they may be represented in completely incompatible ways. It is undesirable to compel sub-system developers themselves to accomplish this. It is the responsibility rather of those who are involved in solving the original problem, which comprises a number of simpler sub-problems, to define protocols to represent common knowledge shared by individual sub-systems and which is needed for cooperation.

Cooperating through Communication: Individual system elements, which may themselves be autonomous systems, "talk" to each other when they are brought together to solve a common problem. As they communicate, they cooperate, they debate, they disagree, they compromise, finally they succeed (hopefully). It is not surprising for them to fail in the end, perhaps due to inadequate individual capabilities, or regrettably - and perhaps more commonly, when the eventual integration of their individual capabilities has not been carefully studied.

Difficulties of Integration: Unfortunately, individual systems themselves have failed to incorporate cooperation as an essential ingredient, although human engineers may be organised in such a way that distribution of the design problem can be effectively managed through cooperation. As a consequence, these systems cannot be interfaced simply to achieve the desired integration. Some of the reasons can be outlined as follows:

(1) Individual systems are developed by different groups of experts (third parties) and these experts may not be motivated towards eventual integration.

(2) There may be significant incompatibilities between structures of systems.

(3) There may be significant inconsistencies between content of systems.

(4) Different systems may be developed under different programming environments.

(5) Proprietary knowledge accumulated among existing systems may be prevented from being manipulated and retrieved directly.

(6) There are technical problems such as memory limitations, programme efficiency, and so on.

1.4 THE AGENTS APPROACH

It is recognised that a key research issue in the study of the utility of knowledge-based techniques is the characterisation of data structures, to maintain compatibility with existing methodologies and reflect the information handling requirements of designers assisted by computers. The three metaphors used in this research are agents, objects and meetings. Among them, the agents metaphor is the primary data structure. Conceptually, an agent is used to simulate and emulate a human designer who may, or may not, be assisted by a computer system. An agent can also represent an autonomous expert system. An object is used to refer to real-world entities such as gears, shafts, etc. An agent designs an object in a similar sense that a designer designs a component. Computationally, an agent is simply a data structure. It is able to manipulate (store and retrieve) information which enables design. An object is generally a more rudimentary data structure which maintains information of what has been designed.

The design role of agents in cooperating expert systems is to act as representatives for different sub-systems. Sub-systems can be developed separately and possibly in different languages compatible with their specific theories and techniques. Agents, however, are developed in a common language in order to communicate and enable cooperation. A prototype language, which is called AGENTS: An Object-Oriented Prolog, has been developed, and its experimental implementation is utilised here, by combining constructs from both Prolog and Object-Oriented (OO) systems.

Metaphorically, a group of agents meets together to solve a problem cooperatively. Appropriate agents are invited to a meeting, which is organised with one or more predetermined main themes. Those agents invited to the session contribute to the solution of the designated portion of the problem. There are several general strategies by which agents cooperate. For example, individual agents may compete for an opportunity of contributing to the activities of the whole group. Each agent will keep track of the design process, seeking opportunities to contribute on its own initiative whilst remaining available in case its knowledge or expertise is required by some other agent. Once it has successfully competed for a chance to contribute, and is granted an authority, it takes over control of the actions and makes contributions. Within its domain of action, an agent may negotiate with other agents to expedite a specific task. In such a fashion, a solution is built up.

1.5 OBJECTIVES

The primary objective is to develop a generic computer system that supports concurrent engineering, in the field of mechanical design and manufacturing, aiming to achieve high overall quality, low cost, and short lead times. The investigation incorporates the subsequent sub-goals.

(1) To investigate various aspects of complexity of product modelling and design, in order to develop a formalism for its effective and efficient management.

The management of problem complexity is essential to design and modelling of products. It forms the basis for knowledge structuring and process organisation of design. Knowledge and organisation of designers are largely influenced by the complexity of design problems and processes. Designers' ability cannot be effectively modelled, and consequently incorporated into a computer system, unless the complexity is successfully managed. One of the key requirements of complexity management is a composite data structure that incorporates evolving product information from multiple viewpoints. This has been widely identified *(Drake and Fela 1989, and Deitz 1989)*, but progress has not been as rapid as the importance of the problem would indicate. The Leeds SE (Structure Editor) *(Shaw et al 1989)* and the METAMODEL *(Tomiyama et al 1989)* are two of the most successful product data modelling systems for representing products from specification, through design, to manufacture.

(2) To develop an intelligent framework for integrated design to support, simulate and emulate how an effective human mechanical design team works cooperatively.

Teams of human designers are highly parallel. That is, team members are specialised. Specialist designers cooperate to fulfil a design project within a differentiated design organisation. How a team of designers should work is beyond the scope of this research. (Indeed, the model described here is very much of an idealised version of real design management.) But how they actually work in real-world design teams is of special interest to develop a computer supported design framework, since designers use computers to facilitate their activities. Methods developed by individual researchers have all been limited, with prejudicial emphases on certain aspects, techniques or stages. A composite framework is consequently needed, although designers do emphasise some of their decision-making activities. *Smithers et al (1989)* describe a substantial amount of research work accumulated within the Edinburgh group. The need for research on cooperative design problem solving has been recognised and some work has been reported *(Bond 1989, Mayer and Lu 1988)*.

(3) To implement a prototype system which incorporates some of the theories and principles of simultaneous product modelling and cooperative product design based on Distributed Artificial Intelligence.

There have been a number of knowledge-based systems for engineering design, some of which have specialised in the mechanical field. Unfortunately, most general-purpose Knowledge-Based systems available for mechanical design problem solving are limited in their own primary domains and it is hard to relate their main features to integrated design. The need for cooperative design has been identified as a priority. However, there is a lack of special-purpose tools to support cooperative design problem solving. Research on Distributed Artificial Intelligence has resulted in a number of languages which are suitable for cooperative problem solving *(Tokoro and Ishikawa 1988)*. However, there has been little evidence that these systems can support cooperative design problem solving effectively.

(4) To develop an intelligent tutorial and instruction system for the education and training of mechanical design engineers and students in that discipline.

Students intending to specialise in mechanical manufacturing and design have to be trained in product design, for example in transmission box design, machine tool design, cutting tool design, fixture design, etc. through course projects. The authors' own experience of being trained, and subsequently being involved in training, suggests that substantial improvements can be made to existing curricula by introducing intelligent instruction and tutorial systems. A successful expert system does not just have the potential to carry out design, but may also acts as a tutorial system to guide naive users, such as students, to transfer experience which has been assimilated in the system. Wherever possible, the system will justify its decisions and explain its intentions.

(5) To evaluate the methodology of the prototype system in order to indicate directions for future research.

The methodology developed for simultaneous product modelling and cooperative product design will be demonstrated and evaluated using the prototype system. Based on the results, it is expected to indicate some of the directions and issues for future research.

1.6 SCOPE OF ACTIVITIES

1.6.1 The Chosen Domain

As indicated previously, the universe of mechanical design and manufacturing has been decomposed into three major sub-sets: design of products, design of manufacturing systems, and design of manufacturing processes, as shown in figure 1.3. The chosen area for this study is production facilities in general, machine tools in particular. Production facilities have a twofold role in manufacturing:

(1) As a specific kind of mechanical product, facilities such as machine tools themselves are designed, manufactured and assembled in the same manner as other products;
(2) Being elements of manufacturing systems, facilities have a functional role in production.

Because of this, research on modelling and design of production facilities can contribute to the study of integrated design and manufacturing, in other words, product design, system design, and process design.

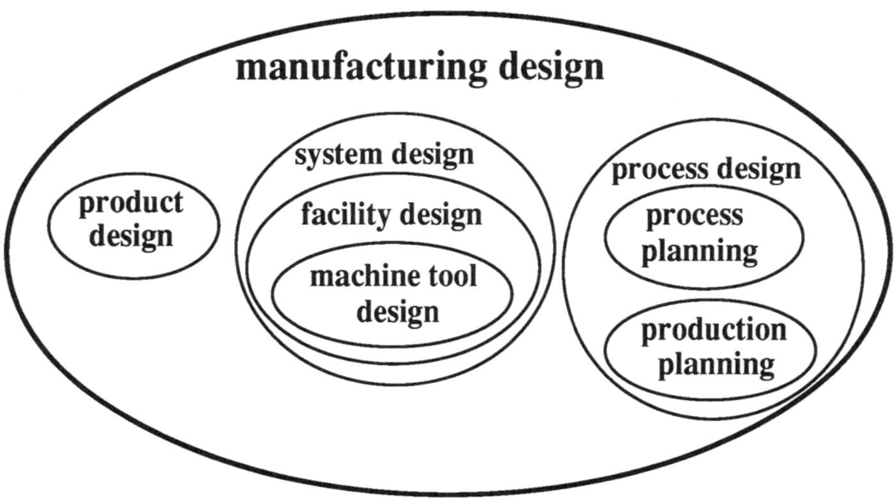

Fig. 1.3 domain of manufacturing design

1.6.2 The Special Interest

Almost all general CAD systems force designers to work at fairly low levels of detail and specificity. During the early stages of a design, designers may not necessarily be concerned with such details as the precise shape and dimensions of a component. Forcing designers to specify additional design or manufacturing details at too early a stage may cause excessive, and unnecessary workloads, but also over-constrain the design, making it converge towards a sub-optimum, limiting options (creativity), or even inhibit the progress of design because of the lack of knowledge with such details.

A central theme of this research concerns the early stages such as conceptual design and embodiment design along the stream of design process *(Pahl and Beitz 1977)*. This is generally regarded as being difficult to achieve:

(1) Design problem solving itself is a very poorly understood activity, especially in its early stages. A simplification is commonly made that only relevant important issues are considered and irrelevant details are omitted. Another simplification is that the product and its components are considered whilst elements within components are, in general, not considered.

(2) The domain of manufacturing design and even its sub-domains are very complicated. A very simple but representative example of conceptual configuration of machine tools is used for illustrative purposes: given a set of descriptions of machining features of a workpiece, the system is expected to generate an overall structure of a machine tool, on which the machining features of the workpiece can be produced. However, underlying methods can be generalised for other types of automation systems, such as a flexible manufacturing cell.

(3) Design activities such as CAD-A and CAD-G are closely related to this research. They are not included, since they are well packaged. However, their utilisation (through interfaces) within a cooperative design environment is essential.

1.6.3 The Technology

The investigation is based on Distributed Artificial Intelligence (DAI) in general, Cooperating Expert Systems (CES) in particular. Both the general science of AI and ES technology have been developed relatively recently. Distributed Artificial Intelligence and Cooperating Expert Systems have only just started receiving attention from the engineering community. Much work remains to be defined for future research. However, AI techniques currently available can be used in design support systems to offer improved performance and enhanced capability. The research reported in this monograph should be considered as a sound starting point for understanding the key issues of cooperative design, although it is by no means complete in the sense of a deliverable turnkey system.

1.6.4 The Programming Environment

The choice of an appropriate computer programming environment is always an early decision for any research project of knowledge-based design. Although Artificial Intelligence is not purely concerned with programming, a good programming environment is necessary to evaluate AI strategies and their practical applications to engineering problem solving.

Ideally, there seems to be a need for a conceptual super-shell which possesses most of the features currently used generally in design systems. In practice, specialised sub-systems have been developed under different environments. Individual knowledge-based systems may have their own specific languages for programming compatible with specialised theoretical approaches and techniques. Interfaces between different knowledge sources play an essential role in cooperation. In this research, it will be argued that different component systems may communicate with each other through common terminology, represented by a common language, implying that a tandem approach may be appropriate.

1.6.5 The Nature of Expertise

Capturing domain knowledge is not an easy task for either knowledge engineers or domain experts, especially in group cooperative design. The nature of knowledge is contextually complicated, from general principles and basic laws to heuristic rules of thumb. The variety of knowledge is diverse, from verbal descriptions to pictorial images, from factual assertions to conditional statements and to inspired speculation. In addition, knowledge is developing. All in all, it must be understood that inadequacies in the knowledge itself, or in its scheme of representation, may exist very often, especially at the beginning of development. Systems with sophisticated mechanisms may be still handicapped by inadequacies of problem solving expertise.

There are not many human experts on all aspects of machine tool design. Rather, their expertise is diverse and specialised. There is, consequently, no existing knowledge readily available for building a computer system. Furthermore, knowledge of machine tool design and manufacturing has traditionally been largely company dependent, although this is gradually changing *(see, for example, Rendeiro 1985)*. One fortunate thing is that some of the tasks in machine tool design are standardised. This standardisation leads to a complex situation. Standard design procedures can be thought of as general knowledge for the system. On the other hand, standardisation forces all the tasks and activities to be constrained and obeyed with few exceptions.

1.6.6 The Context

This research has been application-motivated. It originally started from an investigation into product representation aiming to develop a composite model *(Huang and Brandon 1988b)*. A conceptual description of such a model is reported in an early paper *(Huang and Brandon 1988a)*, with the emphasis on representation of components and relationships between them. This model serves as a basis for integrating activities during the process of design problem solving. A prototypical discussion has been given in *Huang and Brandon (1991)*. Concepts developed in preliminary studies have been extended and improved, in terms of both theoretical development and implementation, by using Distributed AI and Cooperating Knowledge-Based systems approach.

In the literature, a distinction has been drawn between Distributed Artificial Intelligence (DAI) and Parallel Artificial Intelligence (PAI) *(Bond and Gasser 1988)*. This distinction is also maintained in this research. That is, PAI is used to refer to the application of AI techniques in distributed parallel data processing or the application of distributed parallel data processing techniques to improve AI performances. On the other hand, DAI is more concerned with logical aspects of distribution with or without parallelism and distribution in hardware. A similar distinction is maintained between Distributed Expert Systems (DES) and Cooperating Expert Systems (CES) in this research. Distributed expert systems are physically or geographically separated, thus relatively close to PAI. On the other hand, cooperating expert systems are close to DAI, emphasising logical cooperation between constituent systems which may, or may not, be physically separated. It should be pointed out, however, that this distinction is highly artificial. Research in both directions may converge to some extent for mutual benefit.

1.7 SUMMARY

Human engineers typically cooperate through communication to solve complicated engineering problems. In contrast, contemporary expert systems do not. A point has already been reached where human engineers are individually supported by autonomous knowledge-based systems. The concern then becomes the provision of a computational environment which supports concurrent engineering by cooperating expert systems.

Chapter 2
Architecture for Cooperating Expert Systems

2.1 INTRODUCTION

In the field of mechanical design and manufacture, there has been a large number of expert systems reported for gear design, bearing selection, shaft design, and so on *(Chu et al 1987, Fagan 1987, Feng et al 1987, and Hatvancy et al 1987)*. In the main, these systems deal with relatively primitive engineering activities involved in the design of much more complicated products such as transmission boxes. Ideally, an expert system for transmission box design would integrate and build upon these systems, so that proprietary expertise accumulated in existing systems can be employed and exploited economically. Unfortunately, the specifications of these individual small systems are commonly incompatible with the requirements of a concurrent engineering approach. As a consequence, they cannot be simply integrated together to solve larger problems *(Brandon and Huang 1993)*.

The architecture of autonomous expert systems has been well established. It includes three essential components: a knowledge base, a working memory, and an inference engine, and some optional, though desirable, components such as a user interface, an explanation facility, a knowledge acquisition capability, etc. (for more details see chapter 6). However, the general form in which a cooperating expert system should be configured has not been so widely studied. It is of a high priority to develop a framework under which such an integration can be expeditiously achieved. That is, knowledge-based expert systems should be constructed such that they cooperate in a similar way to an expert panel to solve a complex problem.

This chapter discusses architectural issues of cooperating expert systems. The primary objective is to identify an effective approach to systems integration. This should be theoretically sound, practically feasible, and realistically applicable. Initially, some general aspects related to distributed artificial intelligence and cooperating knowledge-based systems are introduced. This is followed by a discussion of various approaches to systems integration. Among these, the tandem architecture is justified, and consequently adopted for further expansion, utilising the agent metaphor. Issues of cooperating expert systems are critically and constructively evaluated, considering items such as communication, collaboration, conflict resolution, and control. Finally, the potential outcomes and applications of the research are outlined with specific references to the context of mechanical engineering.

2.2 DISTRIBUTED AI AND COOPERATING EXPERT SYSTEMS
2.2.1 Distributed Artificial Intelligence

Distributed Artificial Intelligence (DAI) is concerned with coordinating intelligent behaviour among a decentralised collection of autonomous intelligent agents. This involves the ways in which they coordinate their knowledge, goals, skills, and plans jointly to devise solutions of appropriate problems *(Bond and Gasser 1988, and Huhns 1989)*. Agents may vary in complexity from simple processing units to sophisticated entities exhibiting rational behaviour. Agents may work towards a single common global goal, or towards separate individual goals that interact and may well be mutually contradictory. Problem solving is cooperative in that mutual sharing of information is necessary to enable the group as a whole to produce a solution wherever possible. The group of agents is decentralised, in that both control and data are logically, and possibly geographically, distributed. Agents may cooperate at different levels, for instance at the level of dividing and sharing of knowledge about the problem and developing candidate solutions; or at the level of reasoning about the process of coordination among agents.

2.2.2 Cooperating Expert Systems

A cooperating expert system refers to a computer program that can solve, or contribute to the solution of, a problem in the domain of an application, in an analogous way to that in which a human expert or a group of human experts does. Just as a human expert can be assisted by computer programs, a computer expert program can also be assisted by other utility programs. They may provide general-purpose services, or they may be more specialised. A cooperating expert system must be able to coordinate these sub-systems to generate a solution, or range of feasible solutions, to the problem.

A cooperating expert system consists of a number of expert sub-systems as its components such that:

(1) Each component sub-system is autonomous and competent within its own specialised domain; it contains the relevant knowledge, represented in certain formalisms; and may use different techniques based on its own local specific theory for making inferences.

(2) Each sub-system is responsible for some portion or aspect of the primary problem.

(3) These sub-systems may be developed separately from each other.

(4) They can be modified easily at the interface level but not internally to interact, and cooperate with each other, when they are brought together to solve a problem, while they can concentrate on working on their own temporal sub-problems.

(5) Their collective performance can extend the capability of problem solving beyond that which would be apparent from consideration of the capability of the individual component sub-systems.

(6) Individual sub-systems can be added to or deleted from the

group freely without disturbing the performances of others.

This provides a paradigm which resembles the cooperation in manufacturing design analogous to the behaviour of an ideal human design group in order to deal with the distributed nature of problem complexity.

2.2.3 Justifying Distributed AI and Cooperating KB Systems

Some of the reasons for considering distributed artificial intelligence and employing cooperating expert systems can be listed as follows:

(1) A pragmatic motivation for using DAI and CES is that most problems in manufacturing design are inherently distributed in nature. Therefore, the behavioursof both human designers and computer systems are constrained in their organisation so that efficient and effective management can be achieved.

 (2) A positive motivation for considering DAI and CES is that many problems in manufacturing design can be partitioned hierarchically, or in other analogous ways such that DAI techniques can be advantageously applied.

(3) The process of construction of DAI systems can provide insight and understanding about interactions among human organisations, which utilise informal groups, formal committees, task forces and societies in order to solve problems.

(4) DAI can provide a means for interconnecting multiple expert systems that have different, but possibly overlapping or interdependent, expertise, thereby enabling the solution of a problem whose domain transcends that of any single expert system.

(5) The complexity of an expert system increases rapidly as the size and variety of its knowledge base increase. Partitioning the system into a number of sub-systems can reduce complexity significantly. The sub-systems and resultant system are more easy to develop, test, and maintain.

(6) Sub-systems can operate in parallel whenever the problem

under consideration allows. In addition, searching within a smaller sub-system is generally quicker than in a large central knowledge base.

(7) It is easier to identify human experts within narrow domains and subsequently to capture their expertise.

(8) A small sub-expert system can be used as a component of a number of distributed expert systems: expertise and knowledge would not have to be re-implemented.

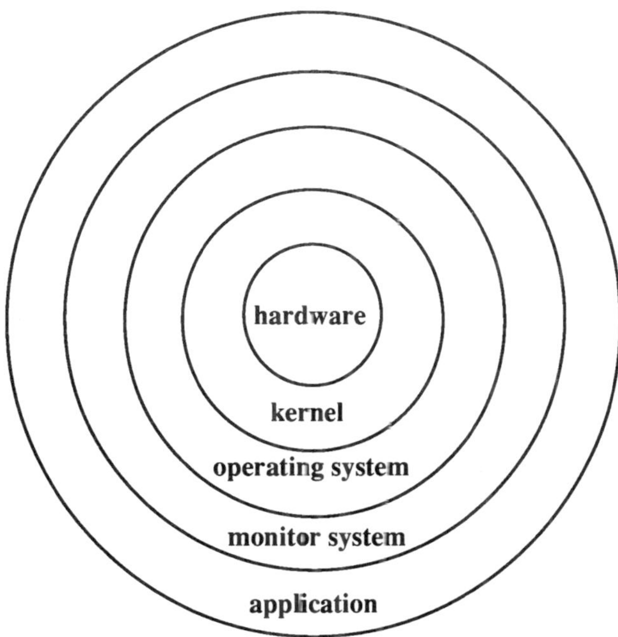

Fig. 2.1 levels of distribution

2.2.4 Distributed Parallel Processing

A distributed parallel processing system can be loosely defined as one that involves the simultaneous operation of multiple interconnected processors. Although this definition is far from being generally accepted, the following four characteristics can be envisaged *(Sharp 1987)*:

(1) Multiple processing units;

(2) Some level of single system image;

(3) Interconnected electronically; and

(4) Significant interaction between units.

However, it must also be emphasised that distributed processing is essentially a logical scheme, which may or may not be mapped on to a physically distributed computer system. Figure 2.1 shows different levels of distribution. Any segmentation at a specific level may result in distribution in the system. Two extremes are distribution at the hardware level and at the application level. In between, it is possible for a system either to run multiple operating systems, as if there were a number of virtual machines, or for a system to operate with different monitors sharing a common base operating system.

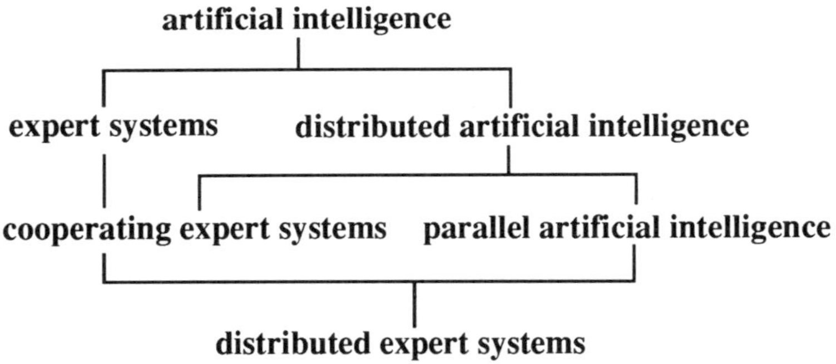

Fig. 2.2 an AI taxonomy

The concepts of Artificial Intelligence have also been applied to analyse distributed processing at different levels. This has commonly been referred to as Distributed Artificial Intelligence. Figure 2.2 shows an appropriate AI taxonomy. Parallel Artificial Intelligence (PAI) is placed under the DAI category to emphasise its primary application in dealing with parallel processing at the lower levels of distribution, especially at the hardware level. In contrast, the particular terms Cooperating Expert Systems or Cooperating Knowledge-Based System are used in this research to indicate distribution at the higher levels. A more specific term Distributed Expert Systems or Distributed Knowledge-Based Systems is used to emphasise the physical or geographical distribution at higher levels including the application level. This artificial distinction is consistent with that usually presented in the DAI literature *(Gasser and Bond 1988b)*, as described in Chapter 1.

It would be contentious to suggest that the discussion on cooperating expert systems in this monograph can be strictly claimed to fall into the category of distributed parallel processing. Conversely, considering distribution at the application level, this research meets the four characteristics of distributed systems, though they may need special interpretation under this circumstance:

(1) There are multiple processing units if expert systems are considered as constituting a class of processor;
(2) The entire system is devoted to a single task of solving a large problem;
(3) Individual systems are connected through effective communication mechanisms; and
(4) The solution of the problem requires significant interaction between participating systems.

2.3 APPROACHES TO SYSTEMS INTEGRATION

2.3.1 Homogeneous Expert Systems

Homogeneous cooperating expert systems share the same Knowledge-Based Management System (KBMS) in terms of knowledge representation, inference, and related activities. They are typical of a design environment where sub-systems are developed under strong central control, using a common 'tool-box' of techniques - perhaps utilising a common expert system 'shell'. Because of the implicit similarities in their structure, the exchange of information and the interaction between constituent sub-systems need only depend on a reasonably simple communication mechanism. However, this approach requires system developers to use the same uniform expert system development shell for all the component systems. In reality, individual expert systems for different tasks have been developed under different programming environments. It may be difficult, therefore, to employ existing expertise directly in knowledge-based systems. Furthermore, the work involved in designing elegant interfaces between systems is still not by any means a trivial activity.

Bearing in mind that traditional expert systems commonly do not have explicit communication structures, there are a number of ways of introducing them. Figure 2.3 shows the three configurations which are most widely used:

(1) In a ring configuration, there may be interacting relationships between individual systems, and therefore interactions must be represented explicitly within the constituent systems. This is often considered not to be feasible because it is not practical to clarify interactions with other systems, which may not even yet exist, at the appropriate stage of development. Another reason is that such a system may not be sufficiently adaptable when it is to be used subsequently for alternative purposes where different interactions with adjacent systems are involved. As a consequence, this configuration is most commonly applied where the number of systems involved is small.

(2) The star configuration is most suitable where there is a substantial degree of similarity among participating systems, because of centralised mechanism for managing communication. This central mechanism is simply a set of procedures which serve as an interpreter between different systems. Every time that a system is to communicate with another system, an appropriate procedure is selected to accomplish the desired interaction. This configuration is, however, analogous to the blackboard integration to be discussed later.

(3) In a hierarchical configuration, communications exist only between systems which have direct links at adjacent levels. There is no freedom for individual systems to interact if they are

not directly connected. In addition, the hierarchical organisation is usually static. Some early applications of Distributed Artificial Intelligence in mechanical design were organised in this way *(Brown 1984)*.

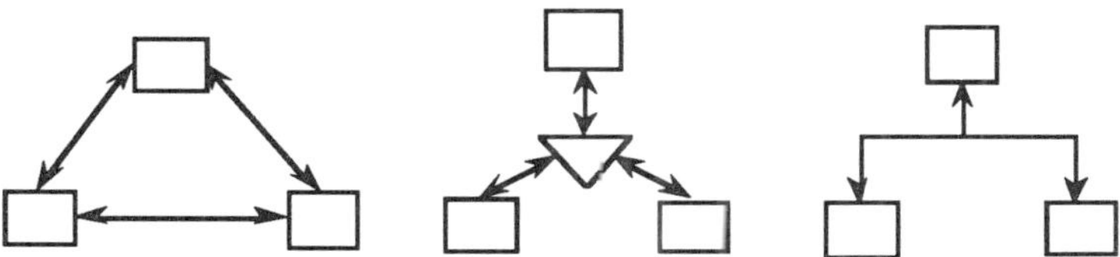

Fig. 2.3 homogeneously cooperating expert systems

2.3.2 Heterogeneous Expert Systems

Heterogeneously cooperating expert systems have differing knowledge-based management systems (KBMS) in terms of their knowledge representation, inference, etc. Because of the differences in their structures, there are difficulties in the exchange of information and interaction between participating systems. For example, two systems may be unable to understand one another simply because the formats in which these systems represent knowledge are different.

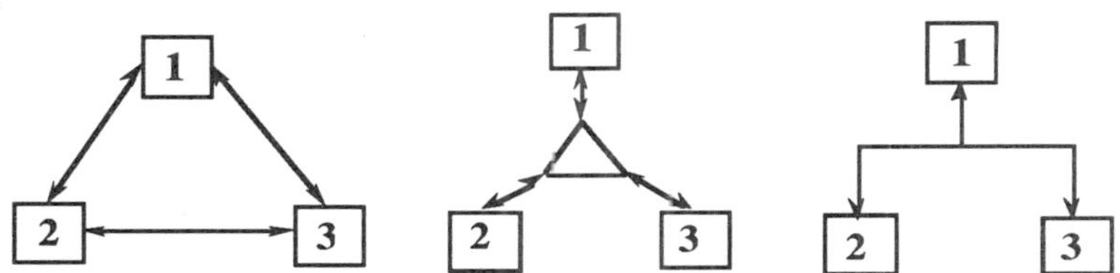

Fig. 2.4 heterogeneously cooperating expert systems

This approach takes a more realistic view since it is able to make use of expertise accumulated from existing knowledge-based systems that might be developed under different programming shells. Difficulties can only be partially overcome by designing a neutral knowledge-based management representation in which knowledge can be manipulated in such a manner that it is understandable and available to all systems. Similar to homogeneously cooperating expert systems, there are three typical ways of organising individual systems, as shown in figure 2.4.

2.3.3 Tandem Expert Systems

As can be seen from the comparisons between homogeneous and heterogeneous systems, described above, there is substantial value in seeking a hybrid system architecture, which retains the global simplicity of homogeneous systems whilst providing the versatility to incorporate the installed expertise and power of existing heterogeneous sub-system components. This may be expedited by use of a tandem architecture. In a tandem architecture a knowledge source consists of two parts as shown in figure 2.5:

- **agents:** The major role of agents is that of enabling and encouraging cooperation. This can be achieved by agents using a common vocabulary, expressed in an argument language, to initiate dialogue and respond to and understand each other, but without the need to know the internal theories, languages and terminologies of other sub-systems.
- **systems:** On the other hand, computation is assumed to be achieved within disparate component systems. Each component system has its own theory involving specialised terminology and techniques, possibly encoded in a specific proof language.

Fig. 2.5 tandem architecture for cooperating expert systems

Generally speaking, the agents may be regarded as application or problem specific, whereas the corresponding participating expert systems are domain specific. A number of agents can be designed to represent a single expert system within different contexts, as indicated in figure 2.6. In consequence, existing knowledge-based systems can be used for cooperative design problem solving without substantial changes, by designing their representative agents carefully. If the expert system has not yet been developed, then its representative agent may be used as a prototype for further development.

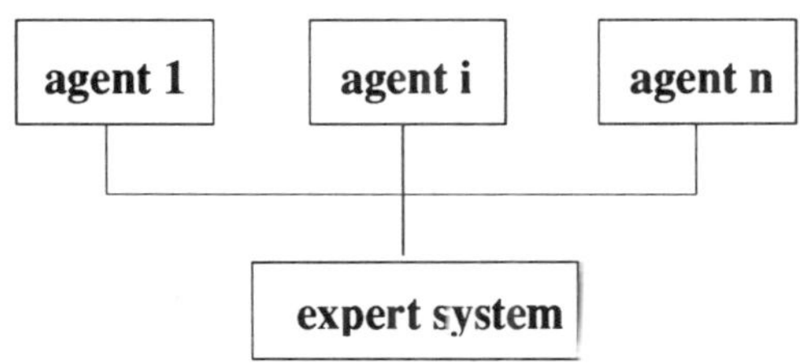

Fig. 2.6 agents and their system

2.3.4 Blackboard Integration

The imagery which lies behind the concept of blackboard systems *(Englemore and Morgan 1988)* is that of a collection of experts congregating round a blackboard in order to solve a problem of common interest, as shown in figure 2.7. Participating experts (agents) interact through the medium of the blackboard by writing on it, and reading from it (cf. message passing between agents). Individual agents may decide for themselves when they can make use of any items in the blackboard and how to behave. In this way, a solution to a problem is gradually built up on the blackboard by cooperative agents.

The basic function of the blackboard is as a data structure, a generally available location for storing common information shared by individual participants. This differs from the central mechanism in the star configurations, described earlier, which is primarily and essentially a set of procedures which translate data into a form intelligible to each participant. Nevertheless, the general forms of these two communication strategies are tending to converge. In particular, some blackboard management systems currently utilise some procedural aspects for purposes such as control. There is also a trend to design individual blackboards for participating agents to retain private information but with a facility for permission to be granted to other agents to read from and even write on them. This is close to direct inter-agent communication of the ring configuration discussed earlier.

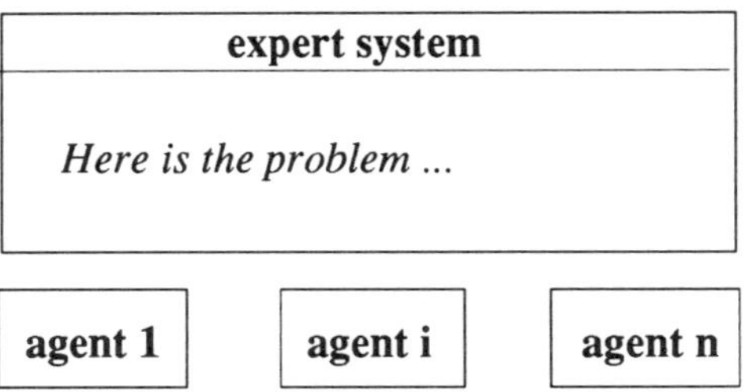

Fig. 2.7 conceptual blackboard structure

2.4 ISSUES OF COOPERATING EXPERT SYSTEMS

Some aspects of the design of of cooperating expert systems can be approached in the same way that is used in conventional expert systems, for instance the methods of knowledge representation or of the inference engine. For some aspects of their design, however, it is more advantageous to exploit ideas derived from systems for distributed problem solving. Since the general objective is to construct an environment which is compatible with prevailing methods of organisation in non-automated and semi-automated design environments, it is necessary to incorporate facilities to emulate the essential features of the interaction between designers and associated functional specialists. In combination, the following are some of the most essential ingredients:

● **Collaboration:** Individual experts must collaborate because of the disparity between their inherent capabilities and developed expertise. The extent of this cooperation can range from totally cooperative to antagonistic, depending on the degree of interdependence of the sub-problems. The disparity implies that firstly they have incomplete knowledge to solve assigned sub-problems in isolation, yet they are required to converge to a final decision; and secondly their skills and expertise are often inexact, nevertheless, they are required to generate a satisfactory solution.

- **Conflict Resolution:** Notwithstanding the need to maintain a general climate of cooperation, it is important that conflicts of opinion between agents should be tolerated, subject to a general requirement to strive for an agreed conclusion (wherever possible). Indeed, local and temporary inconsistency must not only be allowable but is necessary to enable a variety of options to be explored. In general, the integrity and consistency of information and knowledge will improve as problem solving progresses.

- **Control:** Problem solving involves exploration and backtracking in order to propose plans for solutions, to refine the proposed solutions, and to diagnose and remedy failed solutions. In the AGENTS environment both exploration and backtracking are controlled and guided by a knowledge module called the control mechanism. Control of cooperation allows agents to make contributions yet also maintains a high rate of convergence towards a satisfactory solution. A control mechanism may use both domain-independent and domain-dependent and problem-specific knowledge as features of its control rules and strategies.

- **Communication:** The issue of communication is the most fundamental and important. Collaboration, conflict resolution, and control are all achieved through an effective communication between participants. There are two widely used ways that communication takes place: message passing and blackboard. In some sophisticated system such as AGENTS, both of them are provided in different contexts to exploit their particular strengths and to overcome their weaknesses.

The above basic issues will be discussed in detail in chapter 4, with special reference to distributed design problem solving in mechanical engineering. Their implementation within the AGENTS system will be discussed in Chapters 5 and 6.

2.5 AGENT-BASED COOPERATIVE ARCHITECTURE

Figure 2.8 presents a conceptual architecture of cooperating expert systems using the agent metaphor. This agent-based tandem architecture for cooperating knowledge-based systems incorporates the blackboard concept. Agents are executive representatives of individual computing systems which participate in the common discussions in the community.

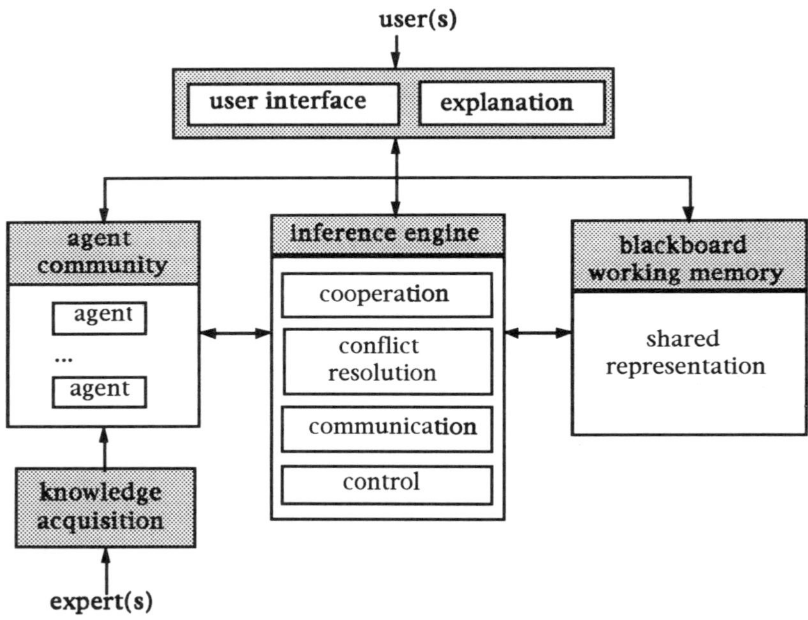

Fig. 2.8 conceptual architecture for cooperating expert systems

The general issues which govern the construction of this architecture of cooperating expert systems may be summarised as follows:

(1) ***Agent Base:*** A cooperating expert system consists of a number of expert sub-systems, represented by agents. An agent can be an expert system (in its conventional sense), a computer program (in its conventional sense, i.e. not expert system) or

even a human expert. Each agent is responsible for some portion or aspect of the primary problem. The collection of all the agents has the capability to span the total complexity of the problem. Individual agents may be developed in isolation or their structure may be inherited from an existing agent. They may be added to, or deleted from, the community freely.

(2) ***Inference Engine:*** The specification for the inference engine of a cooperating expert system must emphasise the aspect of cooperation between agents to propose plans and resolve conflicts. The following alternatives, not necessarily exclusive, are available:

- a single central inference engine shared by all the agents;
- each agent has a local and private inference engine, with no central engine;
- a combination of central and local inference engines.

(3) ***Global Working Memory:*** Whether there should be a globally shared representation of the current state of problem solving by the group of expert agents, or individual agents possess their own local working memories, is a decision confronted by system developers. Currently, some CES systems have a global representation, some have local memories, and some have both. It is desirable, therefore, that a general structure for cooperating expert systems should accept either of these options, including a combination.

(4) ***Knowledge Representation:*** Every agent may incorporate specialised knowledge in its own private knowledge base. For agents to cooperate, however, they must share knowledge and information. One way of sharing knowledge is through a common knowledge base accessible to all of the agents. An alternative method is through inheritance relationships between agents (as is used in object-oriented systems).

(5) ***Knowledge Acquisition:*** Knowledge acquisition in a cooperative environment is a new challenge which is inherited partially from conventional expert systems but with the need to take account of special problems characterised by distribution and cooperation.

(6) ***User/Agent Interface:*** Interfaces between agents are an important issue in cooperating expert systems. The user can be treated simply as a kind of agent within the multi-agent system as a source of knowledge and control.

(7) ***Explanation:*** Individual agents must be able to explain their decisions and intentions to the user. As a group they must be able to explain their joint decisions and intentions to the user. There may also be some kind of explanation for non-user agents.

2.6 POTENTIAL APPLICATIONS IN MANUFACTURING DESIGN

The design of general schemes for intelligent systems for Computer Integrated Manufacture must take into account the huge diversity of problem structures, managerial strategies and skills. These are widely compounded by an industrial culture based on several hundred years of craft demarcation *(Brandon 1992)*. For example, *Waterlow and Monniot (1986)* remark that whereas the overwhelming majority of sub-systems in CIM are developed in a 'top-down' mode (e.g. CAD systems), because the complexity of the technical problems involved transcends the capability of local personnel, Computer-Aided Production Management (CAPM) systems are widely developed from a 'bottom-up' viewpoint, because the technical challenges are likely to prove tractable at a local level.

For these reasons, the opportunities and problems for the development of cooperating expert systems will be considered here in the context of the traditional demarcations of manufacturing systems design. It will be shown that each of the functional groups, corresponding to different aspects of the system design problem, have traditionally developed local product models without consideration of the ways in which dependent functional specialists might benefit from negotiation of a common product model.

2.6.1 Cooperating KB Product Design

One of the motivations of introducing expert systems in product design is that they are expected to solve problems that are difficult to address by purely analytical or geometric approaches. As a consequence, it has been recognised that the expert systems approach is not simply an alternative to the others but is complementary. Early expert systems in product design were constructed for designing simple products. Subsequently, expert systems technology has been applied to reinforce functional divisions within traditional design systems, for example between conceptual design by senior designers and the subsequent detail design by draftsmen with more limited skill and experience. As described by the authors elsewhere *(Brandon and Huang 1993)*, this tends to strengthen the status quo to the detriment of progress towards concurrent engineering. As a consequence, efforts to develop intelligent CAD systems must be understood not only in the context of expert systems but also in the context of the desire to develop a universal scheme for integration of design activities.

There have been a number of expert systems which are concerned with the design of elementary mechanical products such as gears, pulley, shafts, bearings, and so on. These are basic or primitive activities which contribute to the design of a slightly more complicated product such as a transmission box. The central motivation underlying the development of cooperative expert design systems is to employ the knowledge gained from the development and subsequent use of these existing primitive expert systems for elementary components, and wherever possible the systems themselves, to solve a more complex design problem cooperatively. An appropriate strategy can be visualised in terms of the following three stages:

(1) To decompose the product into components and sub-assemblies,
(2) To solve these sub-problems using corresponding expert systems, and
(3) To integrate sub-solutions into larger assemblies and an entire solution.

A second reason for using cooperative problem solving methods is that any product design problem comprises a (large) number of factors and aspects *(Bond 1989 and Ishii et al 1989)*, even when the product itself is quite simple, such as a shaft. For every aspect, there is potentially an expert system that takes its responsibility to ensure that the design is of a high quality from that particular viewpoint. These aspect expert systems interact to ensure that the design of simple entities is satisfactory from an overall viewpoint.

A third reason for employing a variety of design approaches is that a design artifact evolves as design proceeds. For each stage, there may be an expert system which is responsible for a specific topic to generate incomplete but acceptable solutions for further consideration. These systems should invoke one another to generate a complete solution.

Ishii et al (1989) have proposed a framework for simultaneous product design to deal with several product life-cycle issues. However, in that particular work there was a lack of a sufficiently general product model to accommodate information and knowledge relevant to every single issue. The capability of such a framework can thus be extended advantageously by incorporating a comprehensive product data model.

2.6.2 Cooperating KB System Modelling

Traditionally the process of product modelling has been considered in isolation from that of product design despite the fact that the effective representation of product information is essential to successful implementation of an integrated design system *(Ishii et al 1989)* or a Computer Integrated Manufacturing system *(Bloor et al 1988)*. The fundamental issue of product modelling is how best to capture the complex nature of product information using a composite product model. Product design involves a range of multi-disciplinary activities, each of which usually requires its own specialised model, although the same information, possibly with appropriate modifications, may be used in different models. In consequence there is considerable merit in seeking to construct representations which will make design information as widely available as possible without detracting from its local effectiveness.

Early research on product modelling concentrated on geometric aspects. Even within these extremely limited domains, establishing the relationships between (perhaps equivalent) representations of a product is often far from trivial, for example where geometric reasoning is considered, an obvious case being the transformation between wireframe and surface based representations. As described by *Martin (1991)* the mathematics required to provide a general basis for geometric reasoning is complex and far from completely understood. Except in extremely limited and simple forms (in certain proprietary systems), geometric product modelling information cannot be readily transformed so that it may be used directly in analytical models (for example in Finite Element computations); nor can it be readily related to the functional model of the product; nor to its production model; nor *primus inter pares* for early stages of qualitative descriptions about geometry.

The realisation of the insufficiency of current geometric data models for supporting manufacturing leads to the recognition of the desirability of the application of expert systems technology to representational aspects of product modelling.

There is no doubt about the need for a robust and composite product data structure in order to integrate the various aspects of the design process. However, the debate on the feasibility of developing a universal scheme for product modelling continues. A conflicting decision is to utilise the existing limited degree of commonality between models or to neglect it. If the decision is to utilise the commonality, then it is apparent that there are significant incompatibilities between the currently available models for different applications *(for example, see Huang and Brandon 1988a)*. If the decision is to neglect the commonality, then, for every new application, a new scheme must be developed and resulting incompatibilities must be resolved before integration. Currently, the trend is to investigate and develop complex data forms which reflect existing CAD/CAM facilities but have the flexibility to exploit evolving technology.

Various attempts have been made to construct such a model in manufacturing design. The SAM system *(Su 1986)* and the Leeds Product Data Modelling system *(Shaw et al 1989)* are two examples of systems which emphasise the integration of manufacturing data within the product model. METAMODEL *(Tomiyama et al 1989)* is a product modelling system for integrative intelligent interactive CAD systemswhich considers behaviours, structures, and functions.

2.6.3 Cooperating KB Process/Production Planning

There have appeared a large number of research systems for knowledge-based process planning; for example *Wright et al (1991)* listed twenty three systems for CAPP (Computer-Aided Process Planning) alone. AI techniques can be used to enhance part representation, representation of production facilities (machines, tooling, etc.), process planning knowledge, and, of course, making inferences. The vast majority of expert planning systems are rule-based: process planning knowledge is represented by (production or logical) rules, where manufacturing information is represented using appropriately structured databases. One of the problems which has confronted researchers is the representation of part (geometrical and technological) information. This is again dependent on effective strategies for product modelling. Suppose that the part information is taken from a CAD system, an integral component of CIM. However, this CAD part information does not usually include technological information although geometrical information cannot be used directly for process planning since machining features are not explicitly represented. This problem cannot be resolved by either the process planner or product designer alone: they must cooperate to produce a product model which contains product information and generates data for process planning.

Because of its role in integrating a variety of activities of manufacturing design, process planning, in its nature, is distributed and cooperative. The majority of extant expert planning systems, however, do not capture distribution and cooperation. Recently, more sophisticated systems have appeared for intelligent process planning. For example, *Hinde et al (1989)* describe a system called LUMP (Loughborough University Manufacturing Package) for 'design to product' integration. The LUMP system is composed of the following major components:

(1) a conversion system for translating geometric information into manufacturing features
(2) a planning system for generating generic operations for manufacturing a product
(3) a database for emulating manufacturing information (facility information about machines and tools) of a "real" factory
(4) a numerical code generator for producing numerical codes required by machines.

These components are integrated by a truth-maintained blackboard. Although the LUMP system has an architecture for cooperative process planning, it seems that those sub-systems are organised sequentially: one starts to work after another ends, and iterative mechanisms are not indicated in the report. The blackboard does, however, serve as the global reservoir for holding information needed by different sub-systems at different times.

2.6.4 Cooperating KB Layout Configuration

Layout of manufacturing systems, ranging from that of single items of equipment to configuring a total plant, is characterised by the distribution of functions and components *(Hatvancy 1981)*. However, most of the expert configuration systems do not deal with distribution explicitly. Early research in configuration of plant layouts is based on techniques of mathematical programming (operations research). However, a mathematical model is only available after the conceptual design phase, when the overall layout of a manufacturing system has been decided, which does not hold in practice.

One of the crucial issues in configuring assembly of a product is the representation of geometry and spatial relationships among parts. This has been reviewed in an early paper *(Huang and Brandon 1988b)*. Again, a model for representing geometry and spatial relationships is difficult to obtain during the early stages of design when much of the detail is irrelevant and therefore is advantageously omitted. For the reasons suggested, configuration knowledge is usually represented qualitatively rather than quantitatively. Due to the above limitations, the technology of expert systems has been introduced for layout configuration *(Kusiak 1990, and Clarke 1989)*.

The ability to reason about layout in the presence of abstract or incomplete geometry and the capability to partition an assembly in the most appropriate way are of particular interest. One of the most widely used techniques is that of representation of an assembly by a network. The vertices represent parts, and edges represent general relationships. By general relationships is meant that information about how one part is to be connected to another in the assembly. That is, the connectivity information reflects the general aspect of spatial relationships between parts. Symbolic representation of incomplete geometry of parts can now be incorporated into vertices of the assembly network. By applying knowledge about parts and relationships between them, more specific aspects can be established incrementally.

Component connectivity information can be associated with components in a similar sense to that in which a component determines its behavioural primitives. This is the basis for cooperative layout configuration. However, there are few systems which configure a product cooperatively on the basis of component information. *Clarke (1989)* describes a system called PROKERN-XPS for configuring automation systems, which will be reviewed in Chapter 9. The system was not developed specifically for the purposes of cooperative design. However, its sophisticated structure can provide an inspiration to the development of a cooperative layout configuration system.

It is not always clear from which viewpoint the layout of a

product should be configured, in its functional perspective, its geometric perspective, its structural perspective, or all of them. It is also not quite clear at which level of granularity of abstraction the layout of a product should be addressed.

2.7 SUMMARY

This chapter has introduced the basic concepts of distributed artificial intelligence and cooperating expert systems. Four general classes of expert systems have been discussed: homogeneous, heterogeneous, blackboard, and tandem. It has been suggested that an agent-based approach may be used to integrate any combination of such systems. Issues such as communication, collaboration, conflict resolution, and control have been identified as being important in cooperative work. Various potential applications of cooperating expert systems in mechanical engineering have also been highlighted.

Chapter 3
The Management of Design Complexity

3.1 INTRODUCTION

The management of problem complexity in product modelling and design is a recursive process of evolution of problem definition and development of strategies for solution generation. It incorporates the following two primary activities:

(1) **Complexity Reduction:** The complexity of a design problem can be identified and attacked by breaking down the problem into components with simpler complexity, each examining the problem from a different viewpoint. Complexity is reduced by capturing the relationships between, and descriptions about, partitioned components.

(2) **Complexity Representation:** The complexity is expressed, and explicitly represented, for effective and efficient use during the rest of design process.

Complexity management is essential to product modelling and design. By analysing the nature of domain complexity, one can define or identify generic properties of the domain so that these properties can be captured, thereby deciding the feasibility of solving a specific problem, to check the design specification, and to select what characteristics are required. It provides a general scheme from which a design can start, and an adaptive representation for tracing a design history.

The primary objective of this chapter is to investigate a flexible formalism for representing product information. This is achieved by studying how the complexity of design problem solving has traditionally been managed by human designers and how these techniques have been, or have failed to be, incorporated into existing computer-aided-design systems. Such a formalism reflects the operating characteristics of existing CAD facilities but has the flexibility to support subsequent total integration of simultaneous design activities. This is invaluable as a basis for a computer implementation. The emphasis is placed on revealing and managing the distributed nature of problem complexity by cooperating KB design problem solving. The study is carried out primarily using the characteristics of semantic data models *(Peckham and Maryanski 1988)* although some alternative models are of value *(Kawagoe and Managaki 1984, Barbuceanu 1984, Fikes and Kehler 1985)*.

In the next section, various sources of design complexity are explored. Section 3 is concerned with representing domain complexity through appropriate abstractions. Section 4 describes a network representation for domain complexity. In section 5, the evolutionary nature of domain complexity in the course of the design process is discussed. Section 6 is concerned with product life-cycle complexity. The chapter concludes with a section summarizing the formalism for product modelling and design.

3.2 ORIGIN OF DESIGN COMPLEXITY
3.2.1 Relativity of Complexity and Simplicity
Manufacturing design, in general, is a complex human activity. Simplicity or complexity, however, is contextual. For example, it would generally be regarded as a trivial design task to select a bolt for ordinary engineering use: i.e. look up a design handbook for advice on selection of a standard component matching the design requirements. In contrast, designing a manufacturing system is complex because it consists of many components and therefore many factors have to be taken into account *(Clarke 1989)*. However, the design of a bolt can sometimes be a complex task because the specialised domain of bolt design can also be further decomposed into a number of sub-domains such as cap, thread, etc. *(Ulrich and*

Seering 1988). The common viewpoint as to the simplicity of bolt design is based on the fact that bolts are usually considered as primitives in a machine system.

3.2.2 Sources of Design Complexity

There are many different sources of complexity, for example:

- **Multiplicity of design criteria:** A product can be examined in terms of a number of aspects and factors such as cost, functionality, production, maintenance, size, etc. These factors must be considered simultaneously during the process of design.

- **Complexity in number and variety:** A large number and wide variety of components may constitute the membership of the domain and its constituent problems.

- **Properties of Components:** Each component is characterized by a number of properties, some of which are local to the component and some of which are general within the domain. In general, the representation of a component can be regarded as sufficient to determine its properties, though some components can only be fully defined when those of another mutually dependent component are known.

- **Relationships between components:** Components may interact in a non-simple way in terms of their properties. A system is not simply the sum of its components, in that it is a non-trivial matter to deduce the properties of the entire system from those of its component parts. Components in a system are seldom completely independent though they are only loosely coupled in many circumstances. However, when sub-systems interact strongly with one another, complexity increases even though relationships can be captured.

- **Recursive complexity of domain components:** Individual component sub-systems can have similar complexity to their parent system. In general, the complexity of a component contributes to the overall complexity of the system. However, when a system and one of its components are considered during the process of design, the complexity of the component may be greater than is necessary for its eventual representation in the composite system.

- ***Unknown parameters and uncertainty:*** The number of variables in a system contributes significantly to the complexity. The simplest one-one case has only two variables: one independent and the other dependent. There are other cases such as one-many, many-one and many-many. In general, as the number of variables in a system increases then so does the degree of complexity. However, this does not always hold since, if a many-many system is linear, it may be straightforward to solve; conversely, if a one-many system has a high order of relational functions between variables, it can be extremely difficult to analyse.

3.2.3 Domain Complexity and Process Complexity

The above sources of design complexity can be partitioned into two broad categories:

(1) ***Domain Complexity:*** This refers to the complexity of the domain in which the current design problem is specified. This typically depicts the complexity of a design problem (domain knowledge).
(2) ***Process Complexity:*** This refers to the complexity of the process of solving the current design problem, in particular the state of design knowledge.

This distinction is made intuitively for simplicity of analysis, and can also be justified by the tendency for designers to separate the specification of design problems from procedural activities (in fact, design specification is only one of several stages). These two kinds of complexity are closely related. Their management is of equal significance in terms of combinatorial complexity: a large number of steps within a simple domain or a lesser number of steps with wide choice at each step can lead to substantial complexity. For instance, suppose that a design problem can be formulated as 3 steps and each step has 5 choices. The number of total possible results is:

5 * 5 * 5 = 125.

If the same problem can also be formulated as 5 steps and each step has three choices, the total possibility is

3 * 3 * 3 * 3 * 3 = 243.

The idea of this example is not to demonstrate which formulation is superior but to illustrate that both domain complexity and process complexity can be equally significant. Thus attempts to reduce either of them can be valuable.

3.2.4 Complexity and Uncertainty

Uncertainty and complexity are twin problems involved in most design problems. Design uncertainty can arise from a number of sources:

- ***Uncertainty about the design objectives:*** At the beginning and early stages of design, designers do not have a fixed set of objectives. Initial design specifications and constraints serve only as rough guidelines.
- ***Uncertainty about the model of an artifact:*** An artifact can usually be represented as a mathematical model for simulation. Assumptions are usually necessary.
- ***Imprecision and incompleteness (ignorance) of information:*** In the early stages of design, designers do not usually have information in sufficient detail.

Human designers have difficulties in dealing with problems of complexity and uncertainty. Therefore, they should not leave all the jobs to computer systems. As a decision maker, a Knowledge-Based design system must have the capability for uncertainty management, perhaps complemented by human designers. The strategies and techniques, used in diagnostic expert systems are difficult to adapt to a form suitable for design problem solving. Different formulations of the same problem may lead to different degrees of uncertainty and complexity. As a consequence, such a system should support a variety of schemes for formulating the problem, and have the ability of choosing a proper (acceptable or optimal) strtategy in order to minimise the uncertainty and complexity.

Traditionally, uncertainty is managed through an iterative process of decision-making activities. Complexity is managed through the decomposition of a design problem into discrete and manageable sub-problems for parallel or sequential processing. Problem decomposition is employed initially and followed by task segmentation and allocation until all remaining problem elements can be managed and solved with available resources.

The primary emphasis shifts as the design proceeds. In the early stage, the focus is the reduction of uncertainty. Uncertainty leads to significant difficulties in the decomposition of a task and the allocation of tasks to specialised design teams. If the uncertainty is the primary characteristic of the domain, then the integrative processes are relatively ineffective. Consequently, the task team should comprise a relatively small number of generic agents. If complexity is the primary characteristic, then problem solving can be managed effectively in a larger, more differentiated, and more specialised team.

3.3 ABSTRACTION OF DOMAIN COMPLEXITY

3.3.1 Divide and Conquer

It is a common and natural practice to attempt to reduce the complexity of a problem by decomposing it into sub-problems. This is referred to as the process of division. The complementary process of capturing descriptions of properties and relationships between problem components is called the process of conquering. This process proceeds until kernel (primitive or simple) components appear. These are defined as those sub-problems which need not be partitioned further because they are already manageable or those sub-problems which cannot be partitioned further due to (perhaps temporary) limitations on the availability of information.

It is, perhaps, difficult to appreciate the foregoing ideas in the abstract. For this reason, the basic concepts will be examined in terms of the elements of manufacturing system design. Consider, for example, the strategy for the design analysis of a thread-cutting lathe. The divide-and-conquer method commonly entails its decomposition into a number of structural modules, as shown in figure 3.1. Individual modules can be described in terms of their principal functions, leading dimensions, etc. The major function of spindle-head gear boxes in general-purpose machine tools is the control of the spindle rotational speed. In the thread-cutting lathe, however, it is necessary to link this to the axial and transverse feed rates of the cutting tool, because of the functional specialism of the machine which, in its turn, derives from the requirements of the class of components to be machined. The quality of a spindle head therefore determines, to a great extent, the quality of the entire machine tool in the sense that it has direct effect on the machining quality of workpieces.

Fig. 3.1 thread-cutting lathe and its components

It will be appreciated that the design and operation of the thread-cutting lathe may well be an element of a wider problem of metal cutting, which itself is a sub-problem of manufacturing system design. Thus what comprises a system and what a sub-system can only be defined contextually within this recursive framework. The divide-and-conquer method is subsequently applied recursively to the constituent modules. For instance, the module of the spindle head comprises a large number of components including gears, brakes, clutches, shafts, etc., as shown in figure 3.2. Even a brake can be further divided into a number of elements.

The objective of problem decomposition is to achieve a reduction of complexity by transforming the general system properties into internal relations among components and describing them explicitly. In general there will be a wide range of options for decomposition, of which some will be much more efficient than others in terms of complexity reduction. Thus, to achieve this effectively, the designer seeks to identify and exploit conceptual coherencies based on advantageous configurations of attributes of the relevant portion of the design, due perhaps to functions, connections, or the existence of some conceptual identity for that group of attributes. Thus the devising of a suitable scheme for partitioning a design problem is a key step to minimise the apparent complexity. In some circumstances, it will be insufficient to employ just one scheme.

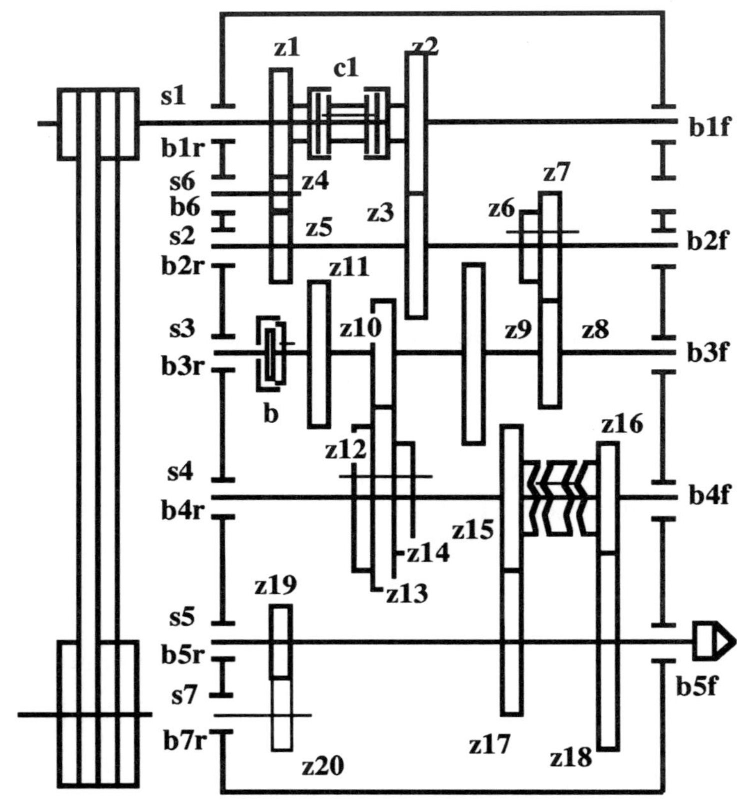

Fig. 3.2 schematic diagram of a spindle head

The rest of this section presents descriptions of a number of abstraction operations for product modelling and design. At present few, if any, systems offer all of these characteristics. Figure 3.3 shows a hierarchy of various abstractions. Set theory is used here for formal definition of the relevant concepts. Graph theory will be used for intuitive illustration.

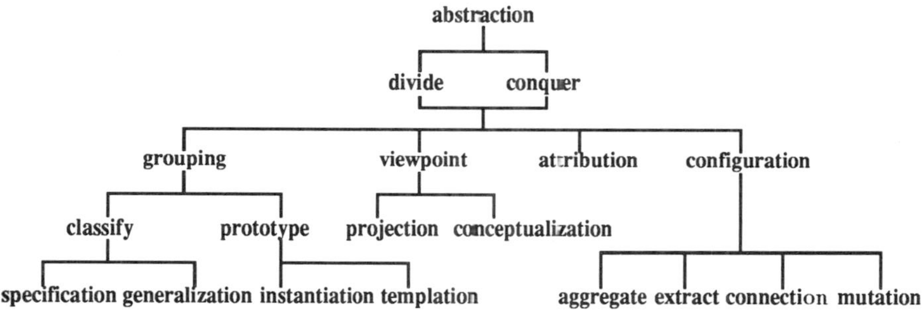

Fig. 3.3 *hierarchy of abstraction*

3.3.2 Attribution

Entities may be real-world objects such as machine tools, gears, etc., or abstract concepts such as forces, motions, etc., or human beings such as design experts. The total set of entities of concern is called the world. An entity is described in terms of its attributes. Attribution is an essential feature of the process of identifying entities and endowing attributes. In set-theoretic terms, attributes can be denoted as follows:

ATTRIBUTE(name, value)
ATTRIBUTE(name, _)

Entities can be denoted as one of the following

ENTITY(name, {attributes})
ENTITY(name, {attributes, ...})
ENTITY(name, {attributes | _})
ENTITY(name, {attribute$_i$ | i = 1, 2, ..., m$_i$})

The world can be denoted as one of the following

WORLD(name, {entities})
WORLD(name, {entity, ...})
WORLD(name, {entity|_})
WORLD(name, {entity$_i$ | i = 1, 2, ..., k})

For example, the world of the thread-cutting lathe in figure 3.1 can be represented by a set of entities:

WORLD(LATHE, {ENTITY(HEAD, ...),
ENTITY(BED, ...),
ENTITY(TABLE, ...),
ENTITY(POST, ...)})

The entity of the gear can be described by a set of attributes:

ENTITY(GEAR, {ATTRIBUTE(MODULE, ...),
ATTRIBUTE(TEETH_NUMBER, ...),
ATTRIBUTE(PRESSURE_ANGLE, ...),
ATTRIBUTE(WIDTH, ...),
ATTRIBUTE(DIAMETER, ...)})

Conception, which is also termed formation, is one of the components of the attribution abstraction. It creates an entity with a given name and a set of attributes in the entire world. For instance, a workpiece entity can be created and added into the lathe world:

PART + {attributes} ⇒ ENTITY(PART, {attributes})
WORLD(LATHE,{ENTITY(HEAD,..),   WORLD(LATHE,{ENTITY(HEAD,..),
ENTITY(BED,...),   ENTITY(BED,...),
ENTITY(TABLE,...),   ENTITY(TABLE,...),
ENTITY(POST,...)})   ENTITY(POST,...),
                                      ENTITY(PART,...)})

Identification is a central feature of the attribution abstraction. It partitions the world into two sets of entities which have, or have not, a given set of attributes. It is a worthwhile and common practice to endow each entity with a unique attribute - its name - which distinguishes it from all other entities in the world.

Description is a facility for manipulating attributes. It assigns a set of attributes to describe a given entity. At the time of conception of an entity, it may have only a few attributes describing its key features. Subsequently additional attributes can be incorporated incrementally by further description. For example, the initial set of attributes describing a gear entity can be extended by inserting other attributes such as its material:

```
ENTITY(GEAR, {ATTRIBUTE(MODULE, ...),
              ATTRIBUTE(TEETH_NUMBER, ...),
              ATTRIBUTE(PRESSURE_ANGLE, ...),
              ATTRIBUTE(WIDTH, ...),
              ATTRIBUTE(DIAMETER, ...)})
                   ⇓
ENTITY(GEAR, {ATTRIBUTE(MODULE, ...),
              ATTRIBUTE(TEETH_NUMBER, ...),
              ATTRIBUTE(PRESSURE_ANGLE, ...),
              ATTRIBUTE(WIDTH, ...),
              ATTRIBUTE(DIAMETER, ...),
              ATTRIBUTE(MATERIAL, ...)})
```

The concept of a facet allows the attributes of an entity to be described recursively:

$$\text{ATTRIBUTE(name,\_)} \Rightarrow \text{ATTRIBUTE(name,\{FACET(name,\_),..\})}$$

This allows entities to include partial description of attributes, and helps to preserve the semantic integrity of a system's knowledge base by constraint of facets. A variety of different facets may be attached to the same attribute, each providing a different type of control parameter or characteristic.

A special role of attribution is in dealing with relationships between entities in terms of their attributes. These are referred to as relationships or relations, denoted as

RELATION(name, {entities}, {attributes}).

Relationships could be described by qualitatively symbolic

descriptions or quantitatively numeric equations. There are two types of relationships: internal and external.

The abstraction relation may take special forms depending on the number of entities in the entity set. If there is only one element in the entity set, then the relation is unitary. A unitary relation deals with internal relationships. In this case, the various attributes of a single entity can be related to each other.

External relationships concern the connectedness between different entities defined according to their attributes. The simplest external relationships are binary, where only two entities are involved.

3.3.3 Classification and Instantiation

The process of attribution potentially results in a large number of entities. They may be difficult to manage due to their large number, even though there may be significant similarities between entities, indicated by their sharing common attributes. Abstraction operations of classification and instantiation deal with similarities and differences between entities in terms of their attributes. These may be used to improve the efficiency and clarity of complexity management.

An entity is called a class entity or simply a class if most of its attributes are defined as variables without values. An entity is called an instance entity or simply instance if most of its attributes have been assigned with values. A class may represent an infinite number of instances by assigning different values to attributes. A class can be denoted as follows:

CLASS(name, {ATTRIBUTE(name, _), ...}).

An instance can be denoted as

INSTANCE(name, {ATTRIBUTE(name, value), ...}).

The process of assigning values to attributes is called parameterisation:

ATTRIBUTE(name,_) $\Rightarrow$ ATTRIBUTE(name,value).

The process of parameterising all the attributes of a class is called instantiation. In other words, the operation transforms classes into their corresponding instances:

$$\text{CLASS(name, \{...\})} \quad \Rightarrow \quad \text{INSTANCE(name, \{...\})}.$$

A class can be instantiated into a number of instances, depending on value ranges for its attributes. Figure 3.4 shows an instance hierarchy comprising a class and its instances.

$$\text{CLASS}_i(\text{name, ...})\{\text{INSTANCE}_{i,j}(\text{name, ...}) \mid j = 1, 2, ..., k\}$$

$$
\begin{array}{ccc}
& \text{CLASS}_i(\text{name, ...}) & & \text{templation} \\
\rule{6cm}{0.4pt} & & & \Uparrow \\
| \qquad\qquad | \qquad\qquad | & & & \Downarrow \\
\text{INSTANCE}_{i,1} \quad \text{INSTANCE}_{i,j} \quad \text{INSTANCE}_{i,k} & & & \text{instantiation}
\end{array}
$$

Fig. 3.4 instance hierarchy

The inverse process of instantiation is templation, which transfers an instance to a class by cancelling values from attributes of the instance:

$$\text{INSTANCE(name, \{ATTRIBUTE(name, value), ...\})}$$
$$\Downarrow$$
$$\text{CLASS(name, \{ATTRIBUTE(name, \_), ...\})}$$

By classification, the world is composed of classes of entities:

$$\text{WORLD(name, \{CLASS (entity)} \geq i = 1, 2, ..., m\}).$$

For instance, the transmission world can be represented by a set of classes of components:

$$
\begin{aligned}
\text{WORLD(TRANSMISSION,} &\{\text{CLASS(GEAR,...),} \\
&\text{CLASS(SHAFT,...),} \\
&\text{CLASS(BRAKE,...),} \\
&\text{CLASS(CLUTCH,...),} \\
&\text{CLASS(HOUSING,...),} \\
&\text{CLASS(LUBRICANT,...)}\})
\end{aligned}
$$

In the above representation of the transmission world, it is not necessary to know how many gears or how many clutches are included.

$$CLASS_i(name, ...)\{CLASS_{i,j}(name, ...) \mid j = 1, 2, ..., k\}$$

<table>
<tr><td></td><td>CLASS$_i$</td><td></td><td>generalisation</td></tr>
<tr><td></td><td>|</td><td></td><td>⇑</td></tr>
<tr><td>|</td><td>|</td><td>|</td><td>⇓</td></tr>
<tr><td>CLASS$_{i,1}$</td><td>CLASS$_{i,j}$</td><td>CLASS$_{i,k}$</td><td>specialisation</td></tr>
</table>

Fig. 3.5 class hierarchy

Classification incorporates an inverse pair of abstraction operations: those of generalisation and of specialisation. Classification results in a class hierarchy, as shown in figure 3.5. Generalisation aggregates a set of entities with similar properties and uses this association to allow the extension to include further entities into a new class:

$$\left.\begin{array}{l} name \\ \{attributes\} \\ \{classes\} \end{array}\right\} \Rightarrow CLASS(name, \{attributes\})\{classes\}$$

Specialisation discriminates between entities, within a given class, to establish new classes:

$$\left.\begin{array}{l} name \\ \{attributes\} \\ CLASS(name, similar)\{sub\} \end{array}\right\} \Rightarrow \begin{array}{l} CLASS(name, \{similar\}) \\ \{CLASS(name, \{attributes\}) \mid sub\} \end{array}$$

Considering, for example, the gear entity in the transmission box, there are two types (classes) of gears widely used: helical gears and spur gears. A helical gear is a gear with its teeth distributed around the cylinder spirally. On the other hand, a spur gear has a zero spiral angle. Thus the helical gear is a generalisation of the spur gear and, conversely, the spur gear is a special case of the helical gear.

3.3.4 Viewpoint: Projection and Conceptualisation

The operation of the viewpoint abstraction is essential to simultaneous product modelling and design. It incorporates two major operations: projection and conceptualisation.

Projection is an operation which entails grouping (classifying) of the attributes of an entity, viewing it from a number of perspectives:

CLASS(name, {ATTRIBUTE(name, _), ...})
$$\Downarrow$$
CLASS(name, {IMAGE(name, {ATTRIBUTE(name, _), ...}), ...})

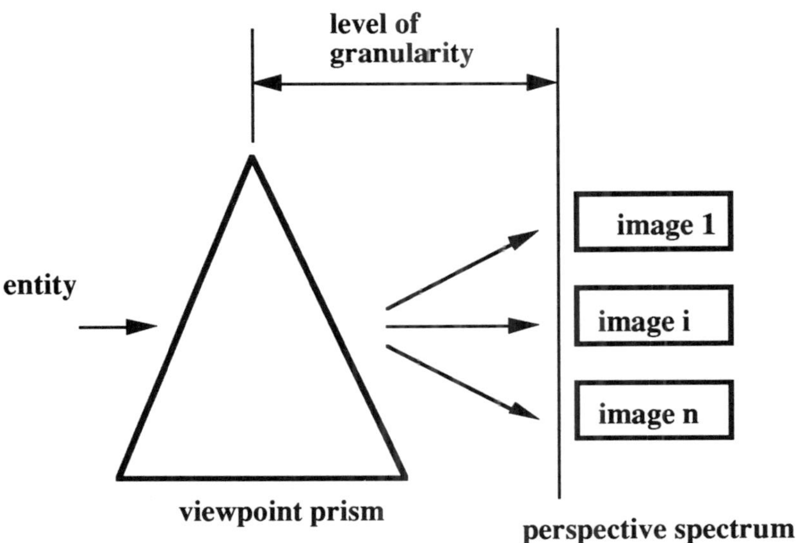

Fig. 3.6 projection spectroscope

A class can be projected onto a number of perspectives to obtain a variety of different images, sufficient to analyse a problem. Figure 3.6 depicts essential ideas of the abstractions of projection and conceptualisation. When an entity passes through a viewpoint mechanism (prism), a number of different images will be obtained of the same entity. Individual images and relationships between images are described in terms of their features, depending on the level of granularity.

Conceptualisation represents a set of images in terms of a set of attributes, thereby relating different features of images from different perspectives:

CONCEPTUALISATION({IMAGE(_, _), ...}, {ATTRIBUTE(_, _), ...}).

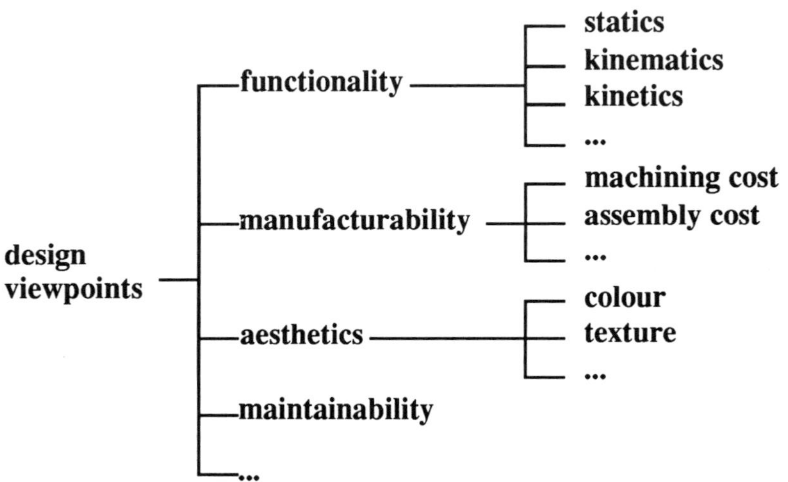

Fig. 3.7 viewpoints in mechanical design

In mechanical engineering, the design viewpoints, as shown in figure 3.7, are often considered. Perhaps, additional factors need be considered depending on the type of product. Within each perspective, the image can be further decomposed in terms of both aspects and components. For example, the size of gear boxes must be considered in terms of axial and radial aspects, both of which are dependent on the sizes of components and their overall arrangement, as shown in figure 3.8.

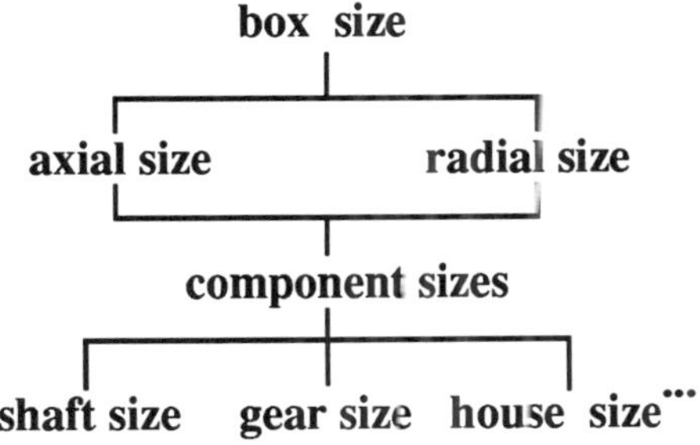

Fig. 3.8 viewpoint hierarchy

3.3.5 Configuration

Configuration establishes relationships, especially spatial ones, between sets of entities, and determines the overall scheme of a world. Two general operations. those of relationship and conceptualisation, have been introduced to deal with relationships between entities in terms of their attributes. Configuration is a special case primarily dealing with spatial and binary relationships.

The essential task of configuration is connection, which relates one component ENTITY with another ENTITY , denoted as

CONNECTION(name, $entity_i$, $entity_j$).

Figure 3.9 shows some examples of typical connections found in transmission boxes.

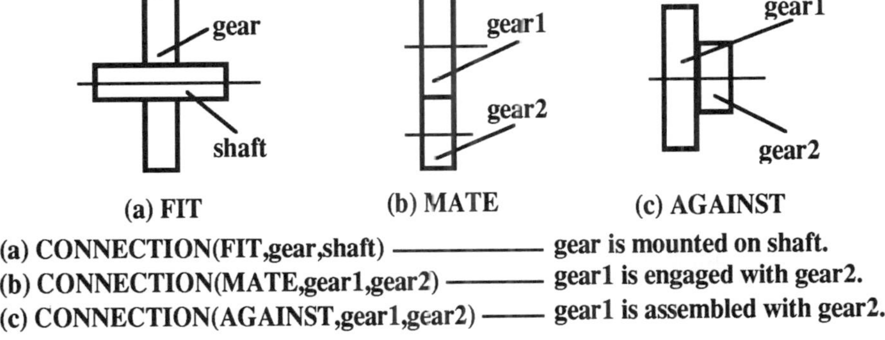

(a) CONNECTION(FIT,gear,shaft) ——————— gear is mounted on shaft.
(b) CONNECTION(MATE,gear1,gear2) ——————— gear1 is engaged with gear2.
(c) CONNECTION(AGAINST,gear1,gear2) ——— gear1 is assembled with gear2.

Fig. 3.9 typical connections in transmission boxes

In isolation, pure configuration assumes that

(1) Properties of entities do not change, or changes in attributes do not affect configuration.

(2) The entities themselves do not change. That is, no new entity is added during the process of configuration;   and   no entity is allowed to be deleted from the set of entities.

It should be noted, however, that the above assumptions do not always hold. In practice new components are added and old ones are deleted from the entity set. In addition, local changes in entity attributes occur frequently. As a consequence it is necessary to introduce additional operations for configuration.

Aggregation defines a relationship between entities (objects) as a higher-level entity (object), allowing details of the relationship to be ignored:

$$\text{WORLD}(\text{name}, \{\text{ENTITY}_1, ..., \underline{\text{ENTITY}_{i,1}, ..., \text{ENTITY}_{i,e}}, ..., \text{ENTITY}_m\})$$

$$\Downarrow$$

$$\text{WORLD}(\text{name}, \{\text{ENTITY}_1, ..., \text{ENTITY}_i, ..., \text{ENTITY}_m\})$$

In this way, a number of related components can be treated as one single entity during the configuration of the world. For example, the transmission entity is treated as one module during the configuration of the overall layout of the lathe world.

Extraction replaces a lumped single entity by multiple entities; it reconsiders previously aggregated entities adding missing elements:

$$\text{WORLD}(\text{name}, \{\text{ENTITY}_1, ..., \text{ENTITY}_i, ..., \text{ENTITY}_m\})$$

$$+$$

$$\text{ENTITY}_i(\text{name}, \ ...)\{\text{ENTITY}_{i,1}, ..., \text{ENTITY}_{i,e}\}$$

$$\Downarrow$$

$$\text{WORLD}(\text{name}, \{\text{ENTITY}_1, ..., \text{ENTITY}_{i,1}, ..., \text{ENTITY}_{i,e}, ..., \text{ENTITY}_m\})$$

$$\text{ENTITY}_i$$

$$\text{WORLD}(\text{name}, \{\text{ENTITY}_1, ..., \text{ENTITY}_i, ..., \text{ENTITY}_m\})$$

Mutation is the action of change in the entities comprising a configuration. In general, toleration of significant variations in the properties of an entity during the configuration process may lead to significant effort although it may be unavoidable in practice.

3.3.6 Granularity, Complexity and Creativity

Domain abstraction results in a necessity to compromise between efficiency and creativity in problem solving. Although creative design, in its complete sense, is out of reach, creativity can be explored in routine parametric design and conceptual design to certain extent. Creativity in knowledge-based design systems is based on the theory of combinatorics *(Ulrich and Seering 1988)*. A wide variety of complex systems can be generated from a small and finite set of elements termed system primitives. These must be clearly defined, described and represented in the database and knowledge base. The manipulation of system primitives may take an unacceptable time to manage by purely mathematical methods, since their complexity increases exponentially with the number of primitives. Furthermore, this process is likely to be extremely inefficient because of the inevitable generation of meaningless results. Therefore, the knowledge guiding the search should also be identified in the domain and represented in the system.

Applying the process of abstraction decribed above, the transmission box used as a practical example can be considered to contain:

(1) a set of class entities,
(2) a set of instance entities,
(3) a number of transmission groups, and
(4) a series of shaft systems.

In an actual design, the above schemes are used in a mixed form. Some intermediate entities such as shaft systems and transmission groups are introduced to reduce complexity and uncertainty.

3.4 NETWORK REPRESENTATIONS
3.4.1 Graph Theory

Many problems in the real world cannot immediately be modelled and manipulated mathematically in a strictly quantitative or numeric form. A scheme for partitioning must be formulated before the specifications of the sub-problems are made. It is by no means unusual to exploit natural partitions of conceptual objects and relations, with much of the detail being irrelevant to the global design at the early stages. Thus, it is common practice to denote a system and its units by vertices, and the relationships between them by edges. This results in a straightforward and intuitive graph which, as the reflection of the real world, is easily understandable. In early papers, suitable topological representation of machine tools has been discussed *(Huang and Brandon 1988a, c)*.

The theory of graphs *(Harary 1969)* is widely used as a general tool for modelling and design *(Harada et al 1978)*. Following the set-theoretic notations of entities and relationships, a graph can be denoted in the form:

$$GRAPH(name, \{vertex_i \mid i=1, 2,...,m\}, \{edge_j \mid j=1,2,...,n\})$$

where

(1) $vertex_i$ has a suitable form for representing entities:

    a label
    ENTITY(name, {attribute, ...})
    CLASS(name, {attribute, ...})
    INSTANCE(name, {(attribute, value), ...})

(2) $edge_j$ has a form compatible with the need for representing the spatial relationships used in configuration:

    a label
    EDGE(name, $vertex_i$, $vertex_j$)

In general, connectivity information of spatial relationships between entities is explicitly represented in a graph (world).

3.4.2 Schematic Networks

A schematic network can be defined as a graph consisting of functional or structural elements (components) and relationships between them *(Ulrich and Seering 1989)*. It is clear that a design represented by a schematic network gives a specification of its constituent functional (structural) components and their interconnections. However, a schematic network differs from other types of network representation in that little of the information concerning detail such as dimensions, geometries, material properties is explicitly represented. This makes it particularly suitable for use at the conceptual stage. Figure 3.10 shows an example of schematic diagram of a transmission box used in a machine tool. With the schematic network representation, conceptual design can be defined simply as the process of generating a schematic network. This is described as synthesis of schematic networks or simply schematic synthesis. This process requires a specification of desired device behaviour as its input and produces a schematic network representing a conceptual design as its output.

3.4.3 Generic Networks

In a specific domain, schematic synthesis does not usually start in isolation. Quite to the contrary, generic components and generic relationships exist in a pre-defined domain. By representing generic components by vertices, and relationships by edges, a generic network is produced. A generic component represents a class of components which share significant similarities in terms of their properties. Therefore, a generic network corresponds to a generic architecture of a design artifact while a schematic network represents a specific architecture. This will be used as a criterion for classifying design problems in Chapter 7. Figure 3.11 shows a generic network of a transmission box. It contains two type of general information:

- Components that usually comprise a transmission box.
- Generic relationships between components.

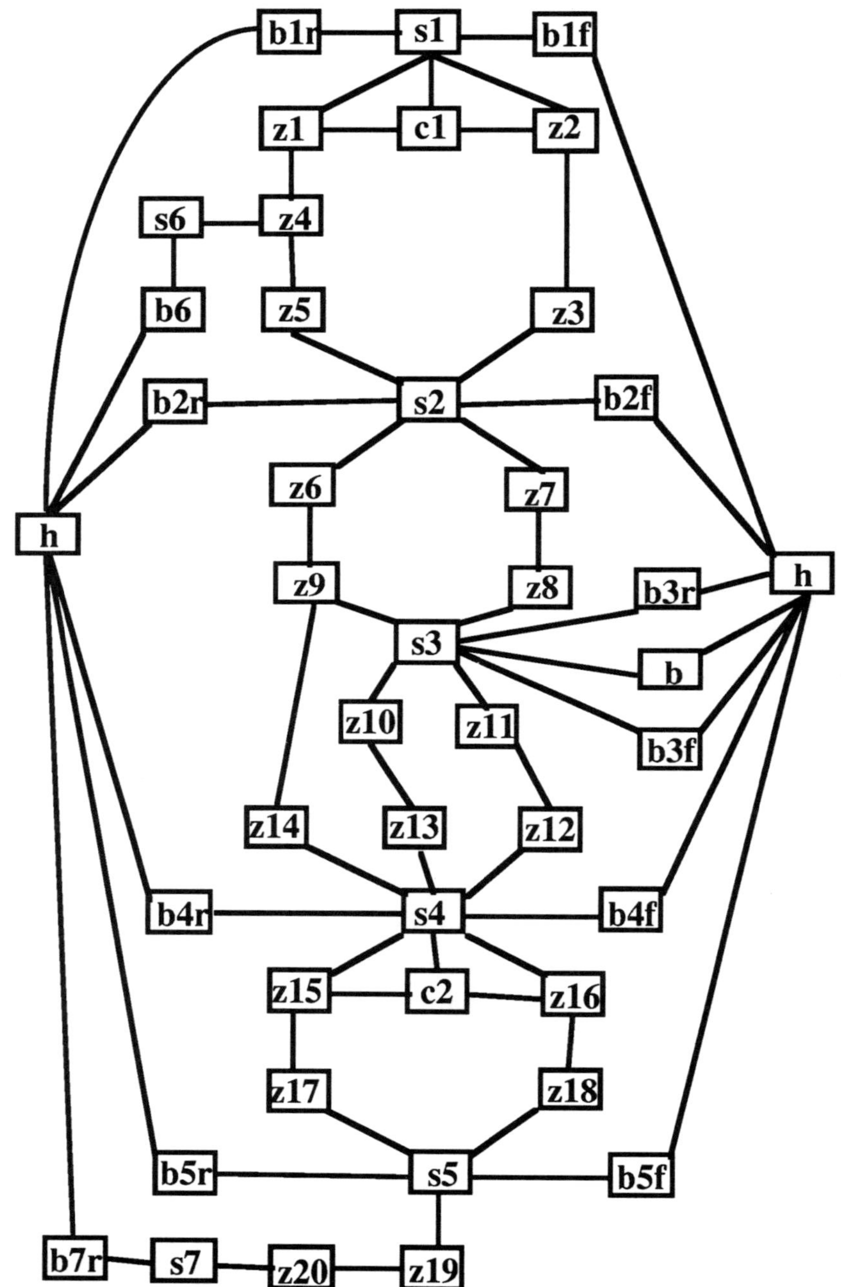

Fig. 3.10 schematic network for transmission box

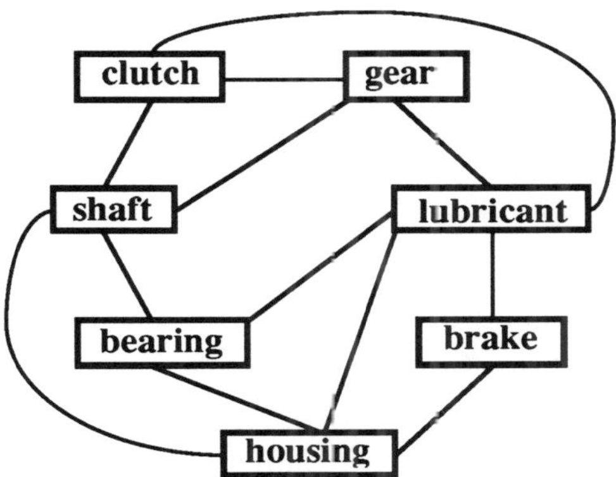

Fig. 3.11 generic network for transmission boxes

A generic network conveys even less information than a schematic network. It does not gives the information about how many components of every class; of course, it does not give information about how components are actually related. For example, how many gears, how many shafts, how many bearings, how gears are mounted on shafts, how a shaft is supported by bearings, etc., are not specified by the schematic network. However, a schematic network is still of great value in representing the generic aspects of problem complexity. The reason is that a schematic network can be obtained if the scope and concern of the domain has been defined. It is not affected by a specific problem. For example, the generic network in figure 3.11 represents a domain of transmission boxes for machine tools, where shafts are parallel and transmission is by means of gears. The generic network represents all the variant schematic networks of transmission boxes by assigning different numbers of instance components to classes of modules, and relating them in different ways.

3.4.4 Characteristics of Network Representations

A graphical representation offers the following two major advantages:

(1) Using an analogous graph helps one to organise data and to exhibit the inter-relationships in question in a simple and

intuitive way. It serves as an infra-structure for stating (with varying degrees of precision) many qualitative notions such as system combination, connections, hierarchies, etc. Based on a graphical model, certain qualitative reasoning and analysis can be performed. This characteristic offers a tool for schematic modelling and design at early stages.

(2) It provides a general structure to accommodate specialised theories such as information theory, control theory, decision theory, etc., minimising the risk of incompatibility, in addition to providing the capability to exploit techniques of built-in graph theory.

3.5 EVOLUTION OF DESIGN COMPLEXITY

3.5.1 Procedures of Design Problem Solving

In general, a problem is solved by executing a sequence of relevant activities through a number of stages. A design activity consists of one or more operations of domain abstraction. A design activity is defined as an action taken

> (1) at a particular time,
> (2) with a particular purpose,
> (3) in respect to a particular portion of the total problem
> (4) using a particular method or technique, and
> (5) requiring particular information and knowledge.

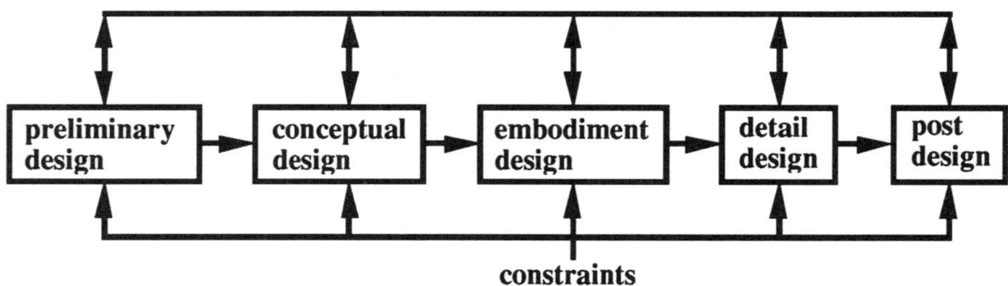

Fig. 3.12 stages of procedural design

Stages in a design process are generally determined by levels of the problem complexity hierarchy. Early stages are concerned with principal requirements and key constraints, while the late stages deal with more detailed parameters. In general, the following five stages can be identified *(Pahl and Beitz 1977)* as shown in figure 3.12:

(1) ***Preliminary Design:*** Any design must entail a stage of requirements processing which is termed here as Preliminary Design. This phase involves the collection of information about general requirements and constraints. Input to preliminary design is very rough and entails diverse requirements, which may be described either symbolically or numerically. The result of preliminary design is the general requirement specification of the product.

(2) ***Conceptual Design:*** The conceptual phase deals with the development of key requirements and constraints to construct an initial state for design. It involves the establishment of functional structures, the search for suitable solution principles, and their combination into structural concepts. This can also be understood as decomposition of functional entities, mapping from the space of functional entities onto the space of structural entities, and layout configurations of entities within the above two spaces. The general framework of the solution to the original problem is determined at this stage. It is extremely difficult, perhaps impossible, to correct fundamental mistakes made at this stage during the subsequent embodiment and detail design phases.

(3) ***Embodiment Design:*** During this phase, design proceeds from qualitative to quantitative solutions. It initiates with the general framework released from the conceptual design phase, further developing the layout and instantiating components in accordance with technical and economic considerations. It is generally not until this phase that analytical tools can be applied to evaluate the design proposals. For example, behavioural performance can be simulated and analysed, financial viability can be assessed, etc. This stage is usually dominated by the configuration of layouts.

(4) **Detail Design:** This phase produces the final solution to the design problem: overall arrangement; forms, dimensions and surface properties of all the individual components are determined; the materials are specified; etc.

(5) **Post Design:** This phase refers to all follow-up activities after a satisfactory design is generated.

3.5.2 Propose-Assess-Revise Cycle

Globally throughout all stages of problem solving, design is a cycle involving three basic activities: propose, assess, and revise, as shown in figure 3.13. This cyclic design process is one in which an approximate solution to a problem is initially proposed and then fed to a mechanism where the proposal is assessed. Based on the decisions from the assessment mechanism, the proposal is passed to the mechanism for revision which reveals a better solution. This procedure is repeated until a solution of a required accuracy is achieved. The overriding principle is that the error decreases with every successive solution.

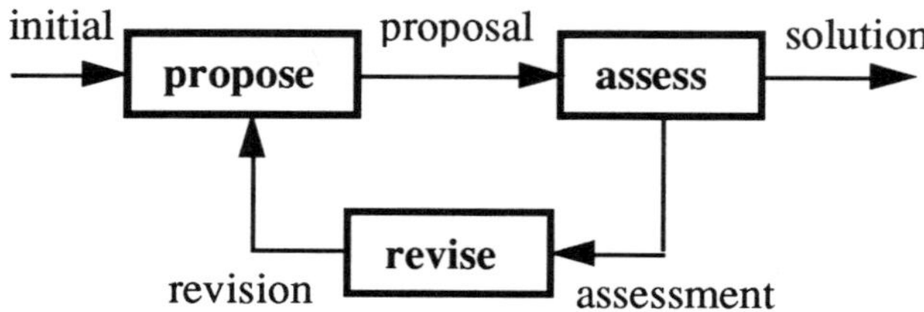

Fig. 3.13 propose-assess-revise cycle

3.5.3 Evolution of Domain Complexity

Understanding of the domain complexity develops gradually as the design process proceeds. The evolution of domain complexity can best be interpreted with the help of network representations. As a result of the preliminary design stage the generic network can be usually obtained. This represents the most general aspects of the problem, i.e.the initial state of the design. As design proceeds, class entities in the generic network are incrementally transformed into instance entities. That is, the network representation is gradually transformed from a generic one to schematic one. During this process, the network is a combination of schematic and generic networks. This network combined with generic and schematic networks are called elaboration networks. Figure 3.14 shows the applicability of various networks during the progress of design. In general, generic networks are useful during the conceptual stage and early phases of the embodiment stage. Schematic networks are gradually built up in the late conceptual stage and early embodiment stage. They can be used as formal models during the latter part of the embodiment stage and during detail stage.

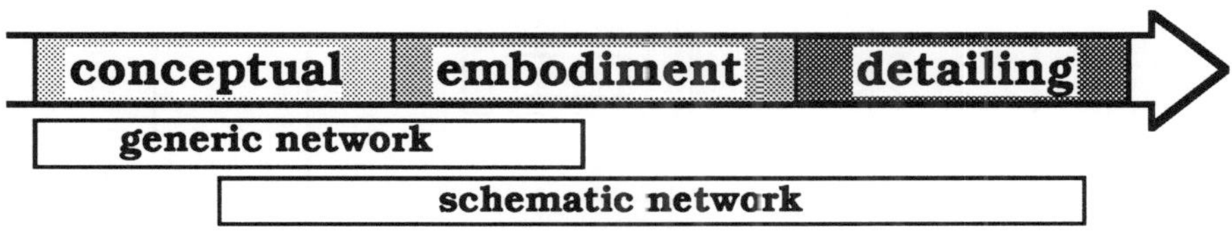

Fig. 3.14 use of networks in design process

Both the initial establishment of either the generic network or the schematic network, and the transformation from the generic network to the schematic network, are evolutionary. Vertices are gradually added and edges incorporated to connect the vertices. Descriptions about vertices and edges are accumulated incrementally from qualitative to quantitative, from general to specific.

Descriptions of an entity, and its relationships to other entities, are collected incrementally and established progressively from the qualitative to the quantitative level of granularity. Qualitative descriptions are primarily symbolic, whereas quantitative descriptions can be both symbolic and numeric. Geometric descriptions, both symbolic or numeric, are in general highly granular. Feature-based descriptions are flexibly granular. A quantitatively correct description must be qualitatively correct whilst quantitative reasoning must be capable of being used to remedy qualitative imprecision.

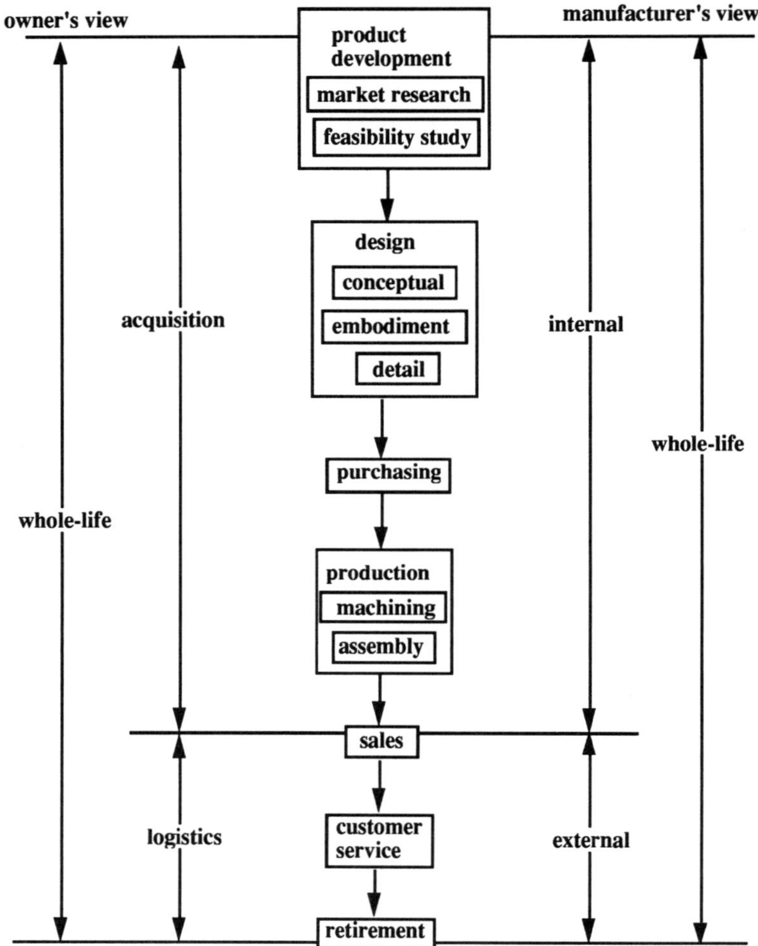

Fig. 3.15 life-cycle complexity

3.6 LIFE-CYCLE COMPLEXITY

3.6.1 Product Life Cycles

Figure 3.15 typifies the extent of whole life of a product and the costs incurred at different stages. It is clear that the whole life cost of a product is the sum of all the costs incurred at different stages along the whole life of the product. However, this summation does not take a simple form because there is a considerable period of time between the product life span, and costs incurred at later stages cannot be compared with those incurred at early stages without considering factors such as inflation. This will undoubtedly complicate the whole-life cost model.

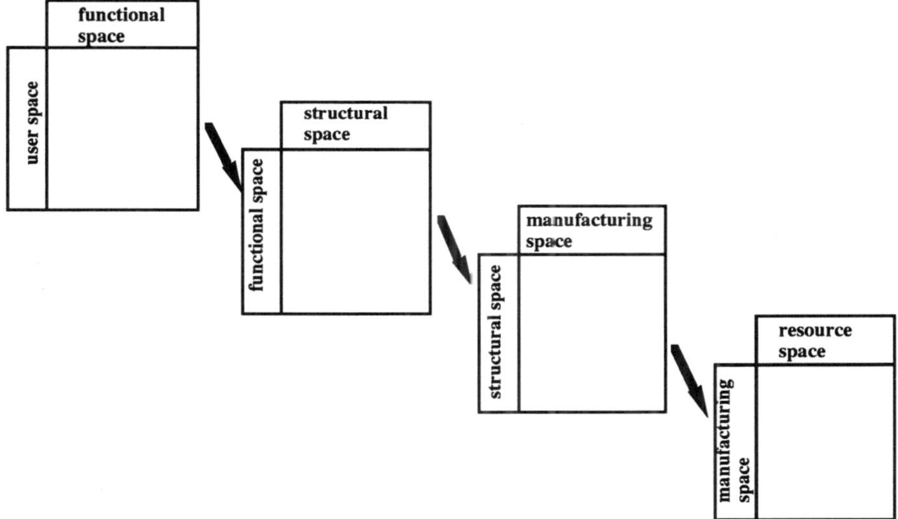

Fig. 3.16 relating requirements to resources

3.6.2 Quality Function Deployment Analysis

Quality Function Deployment (QFD) analysis *(Akao 1990)* is a systematic means of ensuring that the demands of customers and the market place are accurately translated into appropriate technical requirements and actions throughout each stage of product development, that is, to ensure that what is eventually produced meets the original requirements. Basically, QFD analysis uses matrices to relate key factors within different spaces. Figure 3.16 shows the following:

- relationships between user requirements and product functions.
- relationships between functional requirements and structural components.
- relationships between structural components and manufacturing features.
- relationships between manufacturing features and production processes.

3.6.3 Function-Structure-Behaviour Relationships

One of the main roles of conceptual design is to establish the relationships between functions and structures. That is, design can be viewed as the process of mapping functions resulted from the QDF analysis onto the structural space, where detail design takes place and design is specified. However, there is no one-one correspondence between functions and structures. Instead, many-many relationships exist. A single function can be implemented in terms of several structures, individually or in combination; a single structure can provide more than one function. This makes the process of conceptual design extremely complicated to represent in automated systems. Chapter 7 will discuss this further.

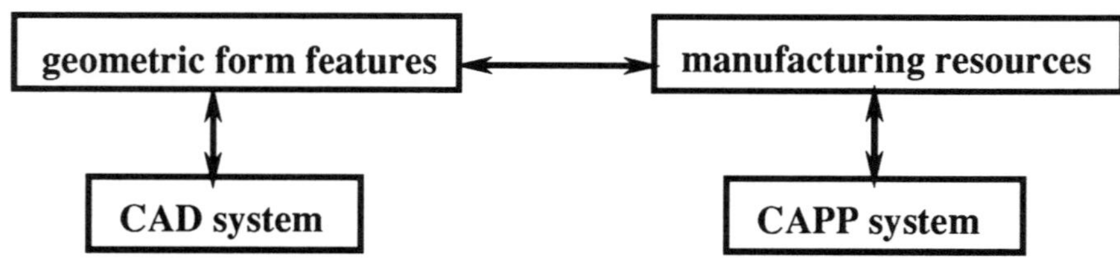

Fig. 3.17 transforming geometric features into manufacturing features

3.6.4 Form-Process Deployment

It has long been recognised that geomerical form features resulting from the design stage cannot be directly used for process planning (in this context CAPP - Computer-Aided Process Planning).

Instead, form features must be first transformed into manufacturing features, as shown in figure 3.17. Significant difficulties have been encountered as far as automation of this transformation process is concerned. Currently, this is commonly accomplished interactively with several proprietary systems being available for this purpose. Alternatively, a feature-based approach is used in CAD systems where design data are represented in the form of features which can be directly manipulated by CAPP systems with minimal modifications.

3.7 SUMMARY

This chapter has presented a qualitative formalism for complexity management in product design and modelling. There is a wide gap between defining the formalism, which is achieved here, and providing complete solutions. This is one of the most important long-term objectives of research programmes in cooperating expert systems. Some of the characteristics of the formalism can be summarised as follows:

(1) The formalism is descriptive: A number of existing data-base management systems fit the formalism. Most existing systems which have been selected as general tools to support manufacturing design utilise only a small sub-set of the universe.

(2) The formalism is generative: With a small number of abstractions, a large number of modelling systems can be created.

(3) The formalism is integrative: A variety of abstractions can be applied at any one time to deal with complicated situations involving heterogeneous views of a product and its components.

(4) The formalism is evolutionary: Abstractions range from symbolic handling to numerical computation, and from qualitative reasoning to quantitative analysis, including evolutionary versions of product descriptions.

Chapter 4
Cooperative Design Problem Solving

4.1 INTRODUCTION

Human design organisations are typically arranged to conform to the distributed nature of design problem solving. However, most computer assisted design systems developed according to traditional methodologies fail to incorporate cooperation as an essential ingredient and, consequently, are not suitable for distributed design problem solving. They are very much limited to relatively narrow domains. Therefore, to solve a design problem of realistic scale, it is necessary to develop cooperative computational design models which are compatible with existing systems. During recent years, a number of attempts have been made to apply distributed artificial intelligence to the development of a cooperative design framework *(Mayer and Lu 1988)*. Brown *(1984)* developed an expert system for mechanical design using a novel approach, differing from traditional expert design systems, being much closer to multiple agent-based distributed design problem solving. Some systems integrate aspects of geometric reasoning, and numeric analysis, into one framework using expert systems technology *(Chieng and Hoeltzel 1987)*. *Ishii et al (1989)* introduced concepts from simultaneous engineering to the domain of mechanical design. However, individual systems have their own specific approaches and it is difficult to extend them to provide a systematic and/or general framework.

The primary objective of this chapter is

(1) To identify common features amongst these existing systems, to extract principles of organising human design teams, and

(2) To develop a computational cooperative design framework which combines generality for theoretical research with a requirement to be sufficiently practical for a prototype implementation.

The study is based on previous design research represented by *Pahl and Beitz (1977), Svensson (1974), and Eder (1980)*, with an emphasis on the element of cooperation in design. It combines ideas from Distributed Artificial Intelligence *(Huhns 1987, Bond and Gasser 1988, Gasser and Huhns 1989, Kornfeld and Hewitt 1981)*, with those from Cooperating Knowledge Based systems *(Fukuda et al 1986, Haren et al 1985, O'Hare 1990)*, Group Decision Making *(Huber 1984, Steeb et al 1984)*, and Distributed Problem Solving *(Decker 1987, Cammarata et al 1983)*.

In the next section, it is argued that cooperation is an essential ingredient in manufacturing design and a realistic computer system would be one that enforces cooperation within distributed expert systems. Section 3 presents a cooperative design model involving agents as individual problem solvers. Section 4 discusses strategies which are used for cooperative design by human designers aiming to develop computational versions.

4.2 COOPERATION IN DESIGN PROBLEM SOLVING

Design is an inter-disciplinary (cooperative) activity involving a group of people with diverse knowledge, perception, resources and experience. This follows from the fact that most design organisations are differentiated and individual members are specialised. However, specific design organisations and their underlying philosophy may vary between companies and evolve with the history of the enterprise. Figure 4.1 typifies some possible departments of a company.

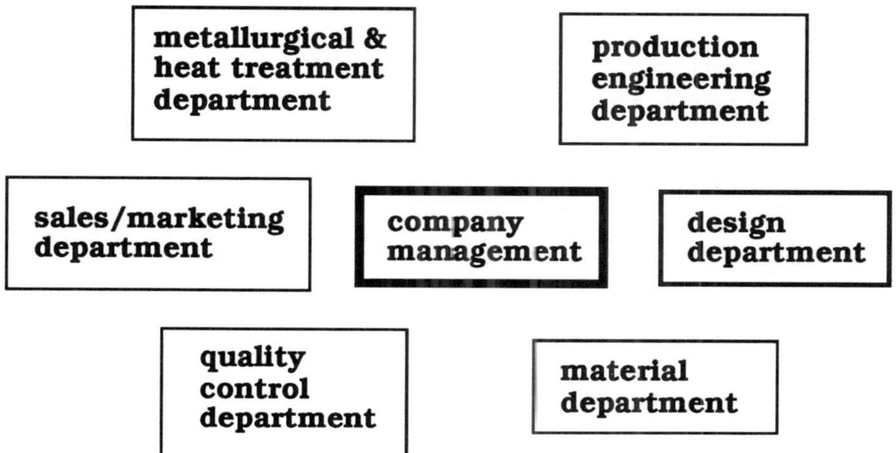

Fig. 4.1 typical departments in a company

A complicated design problem usually cannot be solved by a single designer or specialist since many disciplines are represented. A typical design team may consist of system designers, component designers, and system analysts, depending on the product being developed. For example, a design team for a transmission box may consist of the specialists shown in figure 4.2. The system designer is responsible for the overall design project, making use of other specialists each responsible for designing particular components. The system analyst is responsible for making sure that the resulting system will work.

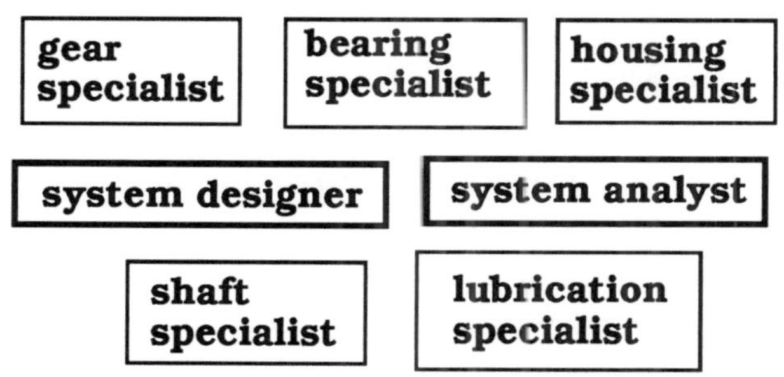

Fig. 4.2 team for designing transmission gear box

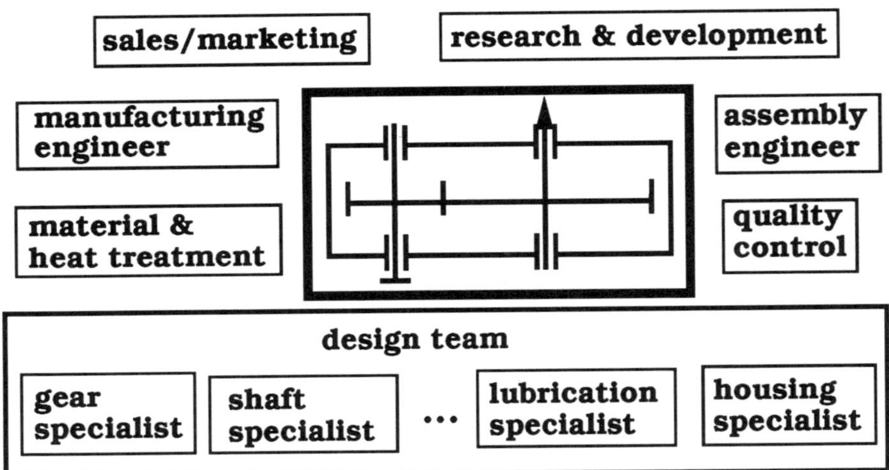

Fig. 4.3 round-table view of cooperative design

It is by no means unusual for designers to fail to define all the necessary factors thoroughly and systematically before going through lengthy design iterations. Nowadays, it is becoming more common for companies to organise their design teams on a project by gathering not only design engineers, but also experts from production, purchasing, marketing, quality control, material, etc., such that they could reach a balanced solution. Although this structure of design organisation is the natural state in the entrepreneurial company, many mature enterprises become too dependent on functional organisation and must restore interdisciplinary organisation through programmes of 'Concurrent Engineering'. For a more specific discussion of the role of Cooperating Expert Systems within a Concurrent Engineering environment, see *Brandon and Huang (1993)*. Figure 4.3 shows a round-table view of cooperative design in the transmission box example.

Success or failure of a design project is largely determined by the effectiveness of communication among members of the multi-disciplinary organisation. Thus inter- disciplinary exchange of ideas is essential. For example, it is particularly important to maintain good interaction during early discussions between design and production engineers. This will result in benefits such as cost savings for example by changing the appropriate component shape (without compromising fitness for purpose) to optimise production techniques, to ensure manufacturability, and to eliminate unnecessary machining.

Specialists outside the design team need not always join every design session if, instead, they impose constraints which must be satisfied at appropriate decision steps. For example, company finance imposes economic constraints on project investment; purchasing imposes constraints on sourcing of materials and components; shipping imposes constraints on packaging; quality imposes constraints on material acceptability, part tolerance, performance envelope etc.

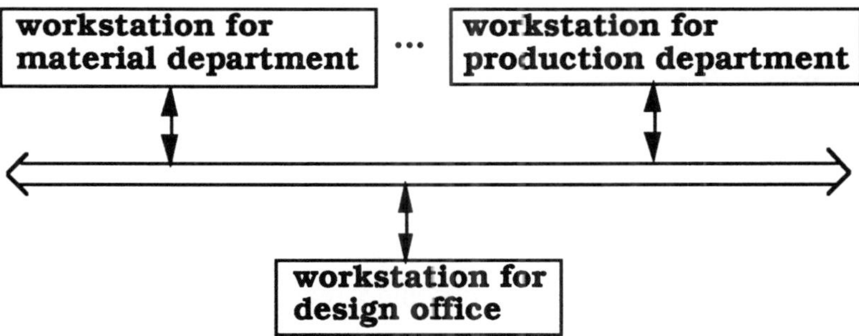

Fig. 4.4 design within a CIM environment

Two potential products of the research in cooperating expert systems in manufacturing design are:

● Design within the CIM network – A distributed knowledge-based system may be developed which enables collaboration between different engineering departments in a computer integrated manufacturing environment. Each department would have its own work- station with its own knowledge base which it controls and maintains. The set of work stations would form a network, as shown in figure 4.4. As far as product design is concerned, the work station in the design department should serve as the central host.

● Design on a single workstation – Knowledge and expertise from specialist departments of a company and specialist members in the design department is assumed available in a form that can be directly used compatibly at a work station which is assumed to be sufficiently powerful to contribute to the overall design, as shown in figure 4.5.

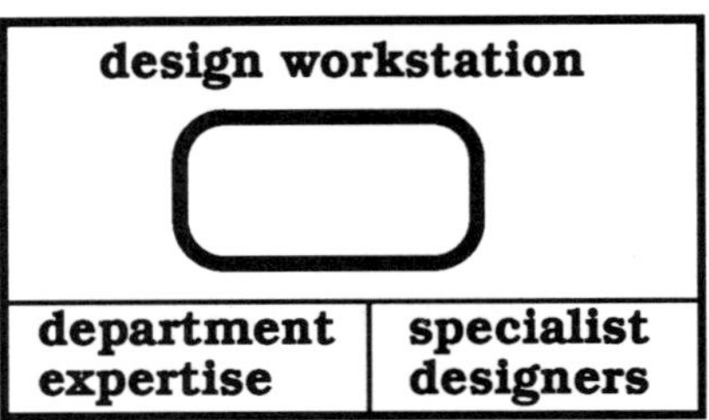

Fig. 4.5 cooperative design on a single workstation

4.3 MODELS OF COOPERATIVE DESIGN

The concept of a design agent has already been used informally in previous discussions to refer to one of the following:

> (1) a human designer or team;
>
> (2) a computer design system/structure;
>
> (3) a computer supported design team (designer).

It is difficult, and arguably unnecessary, to give a formal definition for design agents. Researchers tend to give special names for their particular use even where such constructs are essentially equivalent. For example, *Brown (1984)* uses design agents, *Chieng et al (1987)* use design managers, whilst other names such as knowledge sources, or modules, are also used. In Distributed Artificial Intelligence, concepts such as actors, agents, objects, etc. are utilised. Among those, the term agent is used here. However, no formal definition is given. Instead, several informal understandings are provided:

(1) Design agents are intelligent entities (either human or machine or machine assisted human) which are almost able to solve a problem.

(2) Design agents provide reservoirs for storing information and knowledge. As a nearly-complete problem solver, an agent must have sufficient information and knowledge to be qualified so.

(3) Design agents are data structures in terms of programming. This point will be discussed later when implementational issues are considered.

Metaphorically, a group of agents meet together around a table to solve a problem in the same way that a group of designers hold a meeting or conference. Individual agents make contributions during a meeting which is a design session undertaken by designers and other contributors where appropriate.

The design board may be compared to the physical design drawing board on which human designers work out design plans. It is a common place for agents to represent a problem and record their attitudes towards the problem. In addition, agents may have private design boards used locally.

As a design meeting proceeds, a design board represents the evolving requirement descriptions, design specifications and design history. At the beginning, it may be empty or contain only a few rough design requirements. Contributions are brought into the design board by design agents. This process of accumulation is not linear, in most circumstances, because design involves the exploration of a space of possible solutions. Any statements made by designers are initially considered as assumptions or hypotheses, which may constitute part of a final solution or may be refuted by further actions. Also, design usually involves the investigation of a number of alternative solutions and involves a multiplicity of inconsistent contexts.

The design engine is an inference engine used globally by the group of agents, although it may be advantageous, and hence necessary, for individuals to have their own local inference engines. The main function of the design engine is to generate design proposals. Similarly, there may be a global knowledge base shared by the whole group while each member may have its own local knowledge base.

The Design Knowledge Base contains two types of knowledge:

- ***Domain Knowledge:*** Domain knowledge is used to define the space of possible designs to be explored (perhaps only partially). For example, when designing a transmission box, which is to have a fixed-ratio, single-stage reduction, a decision to use gearing, rather than pulley-belt drive, requires the application of domain knowledge about distances between two shaft axes and other functional requirements.

- ***Design Knowledge:*** Design knowledge is the knowledge about how the space can be explored. This consists of design methods and strategies which are not, in general, restricted in their validity to the specific domain.

It should be emphasised that the above two types of knowledge are roughly divided here for understanding. They are not orthogonal kinds of knowledge: design knowledge in some cases is domain-dependent.

4.3.1 Community of Agents

The agent community refers to the union of all the design agents which may prove relevant to the solution of the current design problem. The strategy for organising problem solving is closely related to the nature of the problem's complexity and its decomposition. Unfortunately, there is no fixed general pattern for relating these factors. In general, the choice of design agents and their organisation is determined by the strategy for decomposing the complexity of the design problem. An effective, and often nearly isomorphic, matching may be achieved by utilising the approximation:

- For each decomposed sub-problem there corresponds a problem solver (an agent).

- For every link between two sub-problems there is a connection between the two corresponding agents.

- For all the implicit relationships between sub-problems there are implicit connections (necessitating negotiation) between the corresponding agents.

Table 4.1 *varieties of agents*

type of agent	characteristics
top-level agent	Top-level agent is responsible for problem solving as a whole. It is the highest but widest to cover all the aspects of solving this particular problem.
upper agent	Agents at upper levels of hierarchy are those in the more general aspects of components.
lower agent	The low levels deal with more specific sub-systems or components.
primitive agent	The leaf level agents solve bottom level sub-problems.
aspect agent	As problem complexity can be decomposed into a number of aspects to obtain corresponding images from different design viewpoints, there will correspond a group of design agents to those problem aspects.
component agent	Within one perspective, design problem can be further decomposed into pieces or components. There may be a set of agents assigned for those components.
general-purpose agent	Some of the design teams, also called service teams, serve general-purpose functions which are common to several of the different activities present in design process. For example geometric modelling and numeric analysis are two abilities that most designers must possess. To solve a specific design problem, there are also teams responsible for special aspects or components of the problem. Geometric and analytical designs are only two generic design activities dominating most of mechanical design. There are others which are less domain independent or less problem-generic, such as cost analysis, functionality, etc.
specific-purpose agent	Any specific design agents are designed for some specific purposes.
user agent	The position of the user in the community of agents is important since the user must deal with multiple agents;some of them are interactive. In general, the user is a problem solving colleague that interacts with other agents and makes contributions to the community; and it is also a form of control.

An agent's knowledge includes which agents it can use and which agent can use it - that is, it knows its position in the conceptual hierarchy, albeit in a restricted way. A typical set of agents contributing to the activities of the design community is listed in table 4.1. Figure 4.6 shows a conceptual architecture of a system for cooperative design problem solving, including components such as the design board, the design engine, the community of agents, and data and knowledge bases, as discussed above. These four components interact with each other, as indicated in the figure.

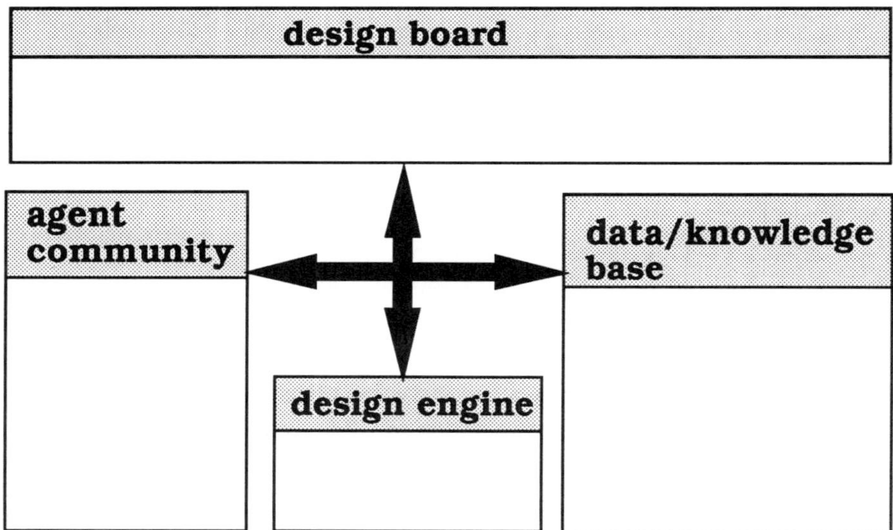

Fig. 4.6 conceptual architecture of cooperative design

4.3.2 Model of One Agent

In a tandem approach, a knowledge source can be denoted by two parts:

- **agent part:** The agent part is essential for cooperation in design. Agents of all knowledge sources conform to a common data structure, so that they are able to understand one another, but without necessarily knowing the internal theories, languages and terminologies of other agents. The commonality basis refers to the terminology shared by knowledge sources in order to cooperate *(Bond 1989)* on the basis of an argument language.

● ***system part:*** Knowledge sources have disparate knowledge. Each has its own theory involving specialised terminology and techniques. This may possibly have been developed in a specialised language, called a proof language. This private proof language may be different for each agent (knowledge source) in that the proof of a statement may depend on different axiomatic bases for individual agents.

Cooperative design is often accomplished by a group of design agents, making their contributions by assertion and retraction of design statements. This approach is also used by *Klein and Lu (1989)*. The following eight categories of statements can be identified:

(1) **Commitments:** A commitment is a task which has been assessed by the group as worthy of exploring, either immediately or later.

(2) **Decisions:** A design decision refers to the result of executing a design commitment.

(3) **Suggestions:** A design suggestion is a tentative design task suggested by an agent but not yet assessed (evaluated) by the group to execute.

(4) **Evaluations:** A design evaluation (assessment) is made by an agent to justify or criticise (i) a commitment, (ii) a decision, and (iii) a suggestion by finding out the associated positive and negative aspects (pros and cons).

(5) **Goals:** A goal statement is issued to describe a task which currently needs to be completed or proved.

(6) **Resolutions:** A resolution statement specifies what to do if a conflict is detected.

(7) **Assumptions:** Assumptions are conditional or unconditional facts for making decisions. However, they may be modified or even refuted in a later action.

(8) **Premises:** Decisions are made on the basis of assumptions and

premises. Premises are facts that hold unconditionally and cannot be refuted subsequently.

Agents play different roles during a meeting. In general, the following five types can be identified:

(1) **The Chair Agent:** A meeting is usually chaired by an agent. This role varies with the strategy by which the meeting is organised. It may or may not be appropriate for the Chair Agent to be a domain expert.

(2) **The Secretary Agent:** This agent is sometimes used during a meeting. Its main task is not that of design but rather to provide a service to the group, for example, writing notes on the design board, maintaining a definitive record of the meeting, etc.

(3) **The Current Agent:** The current design agent refers to the one that is making contributions at present. Here, it is of more interest how and when an agent becomes currently active than how it makes contributions.

(4) **Executive Agents:** An executive agent proposes suggestions, makes decisions, and issues goals.

(5) **Monitoring Agents (also called Advisory Agents):** A critic agent evaluates a design suggestion, a design commitment, or a design decision by identifying positive (pros) and negative (cons) aspects.

4.3.3 Proceedings of Meetings

The state of a meeting is represented by a set of descriptions about entities and their relationships, in terms of their attributes and values on the design board. Agents may change the state of the design board by adding to, deleting from or changing statements. The state of a meeting is characterised by the following factors:

(1) **The Initial Design State:** The initial state of the design, held in the design board, may contain a set of design requirements and a partial set of design specifications. Design requirements may be incomplete in the sense that they fail to identify some necessary

aspects of a possible solution, and they may be inconsistent in the sense that they may specify two or more mutually exclusive properties of a solution or conditions that must be satisfied. On the other hand, design specifications are complete and consistent.

(2) **The Current Design State:** The design meeting proceeds from one state to another. The current state of a design meeting is the previous state plus the changes.

(3) **The Design History:** The entire process of design is recorded as history within the design board. The history of design represents important knowledge gained as a result of carrying out a design task; it represents a record of what regions of the design space were explored, how and why decisions were made; it represents a description of the nature and structure of the space surveyed. The design history is useful during both design exploration and backtracking. Past success may suggest heuristic short-cut routes for exploration, and past failures prevent repetition of earlier mistakes. From the viewpoint of backtracking, design history retains the past partial routes, enabling the design to be returned to some point of a particular route when a failure is detected or a refinement is necessary.

(4) **The Closing Design State:** A design meeting is closed at a success or an irrecoverable failure. A design meeting fails after all the possible alternatives have been exhausted. A meeting succeeds when one or more solution proves acceptable, according to the knowledge available in the system and constraints imposed. After the closing of a design meeting, it is possible to enquire about the solution and how the solution was generated (i.e. explanation).

4.3.4 Conduct of Meetings

Figure 4.7 shows the general process of exploration-backtracking in cooperative design. It can be seen that the data/knowledge base is not explicitly included in this process (c.f. figure 4.6). The explanation is that information and knowledge for design problem solving is distributed implicitly among agents. The global data and knowledge shared by all the agents can be represented by a global agent to which all other agents have access. Elements of the design board and the community of agents have been discussed previously. In the rest of this section, the four functions of the design engine are discussed.

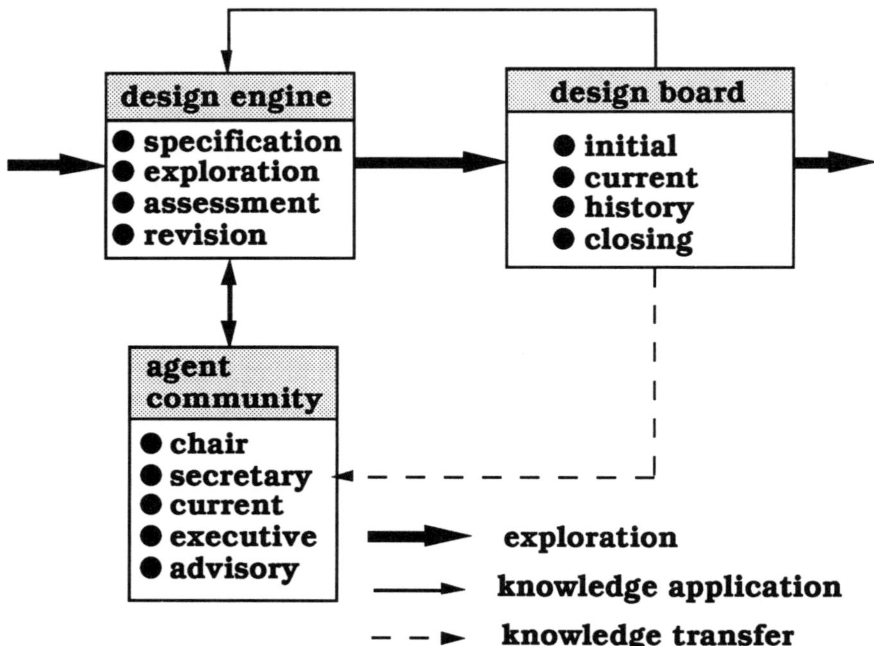

Fig. 4.7 exploration and backtracking in cooperative design

4.3.4.1 Problem Specification

Any design must entail a stage of requirements processing. During this stage the design problem itself is defined, detailed and formulated. Following the definition of input requirements, general strategies of problem solving can be recognised. Problem specification is preparatory so that it is usually undertaken with assistance of human expert designers.

Suppose that there is an agent-based system developed in the manner discussed previously and that it is to be used by a human user to support the processing of design requirements. Input to the preliminary design is very rough and entails diverse requirements, which may be described symbolically and/or numerically. They can be read into the system from a disk file or by interviewing the user. In both cases, they are recorded in an internally equivalent format. Some of the main tasks of requirement processing include:

- **check special requirements:** After all the requirements have been read, the system starts carrying out requirements analysis. First, the requirements have to be checked to ensure that they fall within the scope of the system capability, as represented by every design agent involved in the design team. Any special requirements must be processed, by the user or the system, after acquisition of appropriate specialist knowledge, otherwise an invalid decision will be made. Alternatively the system may be unable to reach a decision at all.

- **classify requirements:** Different design agents are concerned with different categories of requirements. The categorisation can be carried out by posting a relevant requirement to an agent if the relationship "who is concerned with what?" is known in advance. On the other hand, a design agent is able to determine whether it has an interest in a specific requirement, simply by checking whether the requirement is contained in the concern set within its knowledge base. If no record is found in its concern set, the system assumes that the design agent is indifferent to this requirement, that is, this requirement has no effect on its decision making.

- **detect interrelations between requirements:** A particular requirement may be of concern to several agents. The effects of two or more requirements when they appear concurrently may well be, and commonly are, different from those when they are considered individually. Interactions (interdependence and conflicts) are involved. For a specific agent, two requirements may seem to be related, but for another agent they may not interact at all. In dealing with requirements interactions, cooperation between the agents is usually necessary.

- ***represent requirements:*** Identified requirements are described or modelled formally and expressively.

4.3.4.2 Proposal Generation

The design engine progressively explores the possible design under the constraint of available resources. During the design exploration incompleteness and inconsistency of design requirements are revealed and resolved, new design requirements evolve dynamically, and finally they are transformed into a set of complete and consistent design specifications. The stages of the process are:

- ***Decomposition:*** in which the requirements, and thus the exploration task, are broken down into more manageable sub-problems and tasks. This requires knowledge of requirements, i.e. the current requirement description, and generates knowledge used to identify other activities and any precedence constraints on them.

- ***Strategy planning:*** in which profitable orderings of intermediate design goals and sub-problems are identified, with a view to the early determination of significant design features, and the identification of difficulties. This requires knowledge about task decomposition and generates knowledge about which activities are usefully executed and in what order.

- ***Scheme generation:*** in which possible solution or partial solutions to sub-problems are formulated. This requires knowledge of the associated requirements and other constraints imposed by the decomposition. It generates initial design descriptions that can be developed in detail and put together to form the overall specification.

4.3.4.3 Proposal Assessment

There may be a number of proposals generated for a problem. Each of the proposals must be assessed: both its intrinsic value and whether it is worthy of further consideration. A design proposal can be assessed in the following three ways:

- history of design decisions
- constituent components
- both of the above

A design proposal is generated through a sequence of decision making actions. Every decision step can have effects on the quality of the eventual solution. This implies that the total value of a proposal can be aggregated from those of individual decision steps. This strategy can only be used when a design history is maintained during the process of design.

A design proposal can also be represented by its constituent components and their relationships. Every component contributes to its total value. Therefore, the quality of a design proposal can be evaluated by assessing its individual components.

Whenever a group of experts gathers, each has its own viewpoint and impressions relating to the entity (what is to be designed). Experts from different viewpoints put individual emphasis in expressing their attitude, towards a design solution. There are two ways that experts can assess a design proposal:

(1) Experts work together to assess a design proposal by studying either its decision steps or its constituent components, or both, to obtain a consensus opinion about a single decision step or component.

(2) Experts work alone to assess a design proposal individually to develop their own opinions. Subsequently the experts then combine these opinions to obtain a composite consensus opinion.

Proposal assessment may include the following activities:

- Assessment of a design proposal to attain agreed objectives. In general an analytical approach is used to evaluate a design proposal on the basis of an objective function.
- Comparison of the assessed objective of a design solution with the criterion of acceptability. The acceptability can be interpreted in a number of ways such as feasibility or optimality, depending on

the quality of knowledge incorporated.
- Decisions as to whether to accept the design proposal or to perform some form of redesign.
- Recognition of deficiencies, i.e. incompleteness is identified and/or inconsistencies are discovered.

4.3.4.4 Design Revision

A design proposal will be revised as a result of the proposal assessment. During the process of revision, requirements are modified, options are elaborated, and, as a consequence, the solution is improved. A decision to revise is dependent on the following factors:

(1) The primary action which leads to this proposal.
(2) The quality of the proposal.
(3) The current topic of concern.

A number of strategies are available, including:

(1) **Failure Handling:** Design is backtracked when one or more requirements or constraints fails to be satisfied.
(2) **Design Iteration:** Design iteration is used to obtain alternative problem solutions by criteria defined in solution evaluation such as possibility, feasibility or acceptability.
(3) **Design Recursion:** Design recursion is applied when the same procedure can be employed repeatedly as a cycle until a solution is found or a failure is met.
(4) **Design Modification:** Design modification improves a design proposal by modifying the values of some of its attributes but without any change in its general form.
(5) **Design Optimisation:** Design optimisation establishes the best design plan or 'tunes' a design proposal to the level of optimality.
(6) **Re-Design:** Re-design must be carried out if no acceptable solutions to a design problem can be found during a design course.

4.4 STRATEGIES FOR COOPERATIVE DESIGN

In this section, a number of strategies for cooperative design problem solving will be discussed. Some have already been implemented in computer assisted cooperative problem solving systems, however some have not yet been attempted. Their applications in group (team) design have been summarised in some published reports *(see for example Svensson 1974)*. More recent views are reflected in *Guide (1984)*.

4.4.1 Communication

There are following seven factors for structuring communications *(Chang 1987)*, but not all of these factors have been taken into account in practical implementations:

sender	<body sending the message>
receiver	<body receiving the message>
content	<what the message is>
timing	<time of passing the message>
medium	<means of passing messages>
protocol	<way of passing messages>

The blackboard and message passing are the two main paradigms most often used as communication media for group decision making *(Decker 1987)*. A blackboard is also a commonly used control mechanism in Knowledge–Based problem solving *(Englemore and Morgan 1988)*. Communication protocols specify how particular messages are passed from the sender (agent sending the message) to the receiver (agent receiving the message). There are two basic protocols widely used for cooperative problem solving: selective and broadcasting. SELECTIVE communication means that messages are targeted to a specific receiver (who is believed to be interested in them). Another selective protocol, that of ON-DEMAND (AT-REQUEST) communication, implies that messages are only sent to a specific receiver at his/her/its request. BROADCASTING communication means that messages are sent to all the agents in the system (no matter whether or not they are interested in them).

When a message is received, the expert checks if it is in the area of interest. If so, it processes it; if not, it leaves it unchanged. Another type of broadcasting protocol is that of passing a message to a blackboard so that all the system agents can read it subsequently.

4.4.2 Strategies for Cooperation

In the following section, a number of strategies widely practised in creative design within a cooperative environment are reviewed with the intention of assessing their suitability for refining for computer implementation.

4.4.2.1 Agenda-Based Meetings

The schedule of a meeting may be determined by an agenda. Contributors may only make contributions according to this agenda. The concept of an agenda has been used as a strategy of control by many autonomous knowledge-based systems *(such as Stefik 1986)*. Here, the agenda is static, i.e., no tasks may be entered onto the agenda during the progress of a meeting. Such a static agenda is often called check-list. The agenda can be used to define:

- who speaks at a meeting and
- when the speech should start.

This is suitable for the following two situations:

(1) When it is important that certain agents should not be overlooked.
(2) When it is possible for a solution to be generated by executing tasks prescribed in the agenda.

4.4.2.2 Brainstorming Meetings

Brainstorming is a technique devised for assisting creative thinking within a group of people with some prior knowledge or expertise which relates to the problem *(Osborn 1963)*. It provides a free-for-all style of cooperation: every agent does what it can do to the problem. Some of the basic rules concerning the arrangement of a brainstorming meeting are summarised as follows:

(1) Define the problem to be solved.

(2) Participants make contributions in turn, but without raising criticisms or passing judgments.

(3) Ideas must be recorded.

(4) Combination and development of ideas are encouraged but care must be taken to ensure that criticism does not occur.

(5) Every item is considered in turn on its merits, moving onto the next item only when it is considered that ideas have been exhausted.

After the brainstorming session has exhausted the group, the next step is to cull out one or more useful idea for further consideration. The problem with this particular method of problem solving is the amount of irrelevant discussion (noise) which can be generated due to (albeit deliberately chosen) lack of leadership.

4.4.2.3 Contract-Net (Competitive Meetings)

Viewing sub-systems and the user designer as agents anthropomorphically, all of them look at the same centralised blackboard on which the problem is described. Individuals read from the blackboard, carry out their own relevant reasoning, and decide what they can contribute to the community, when this is considered worthwhile by all the members. Individual agents compete for the opportunity to contribute to the activities of the whole group. Each agent notes every opportunity that it believes is helpful. Once it has competed successfully for an opportunity, it takes over the actions. In this way, the solution to the problem is gradually built up by collective actions from cooperative agents.

The above process can be interpreted in terms of a contract-net which requires activities of advertising and bidding *(Smith and Davis 1978)*. The contract net models transfer control in a distributed system using the metaphor of negotiation among autonomous intelligent agents. The net consists of a set of agents that negotiate with one another by means of a set of messages. The agents represent the distributed computing resources to be managed. In any given transaction, three classes of agents may be identified:

(1) The Manager is the agent node that identifies a task to be done, advertises it to other agents for bidding, and assigns it to one of them for execution.
(2) The Bidders are agents that offer to perform a task.
(3) The contractor is a successful bidder, i.e., the one whose bid has been accepted by the manager.

Agents communicate by means of standardised messages. The essentials of contract-net process are summarised as follows:

(1) A manager advertises a 'Task' announcement describing the task to be done and the criteria for bids.
(2) Bidders respond with 'Bids' to announce their willingness and capabilities to perform a task.
(3) The 'Award' message from the manager to the successful bidder establishes the bidder as the contractor for the task.
(4) The contractor sends an 'Acknowledgement' of the award, either accepting or rejecting it.
(5) The contractor sends 'Report' messages to the manager announcing the current status or termination of a task.
(6) The Manager may send a 'Termination' message to a contractor to interrupt its performance of a contract prematurely.
(7) Idle agents may broadcast their availability with an 'Availability' announcement.

4.4.2.4 Synectic Meetings

'Synectics' represents another approach to group creative thinking, similar to the brainstorming method *(Gordon 1961)*. In this case, however, the organisation of the group is more structured: members of the group must have special knowledge and skills in the area involved and only the chair agent is aware of the specific problem to be solved. The basic rules of the synectic approach can be summarised as follows:

- The process is under the general control of the chair agent who leads a general discussion on one topic at a time which is believed to be central to the problem. The chair agent may not contribute to the business of the group.
- Participants in the group, made up of varying skills, are given opportunities to select a solution.
- The chair agent provides some guidance and stimulation by questioning and offering significant, but not problem revealing, information.
- The metaphorical role of the chair agent is that of a client with a problem. The solution can be implemented at the end, because the client has sole authority to implement it.
- There is an observer who is trained in synectics and observes the session, and who, afterwards, leads a discussion, reviewing parts of the best three ideas or points, and the worst one.
- At the outset the client describes the problem, indicates the measures already taken, and suggests the areas in which help is expected. Questions are unnecessary; they are in any case discouraged and regarded as a form of noise. During this period participants write down their ideas.
- Concepts are then suggested which, if reliable, might lead to a solution. These are written on a blackboard by the leader.
- To ensure a positive atmosphere, concepts are expressed as "wishes", but they are not evaluated or criticised at this stage.
- The leader then seeks one or other of the ideas from the participants which appears to fulfil the expressed "wishes" of the client. The leader then calls for an idea to meet any objections which are expressed, repeating the process until the client has no further reservation.
- At the leader's discretion, the participants can be invited to expand an idea before the client expresses an opinion. This procedure may produce a solution rapidly, although it cannot be accepted until the client declares that it now satisfies requirements.

4.4.3 Strategies for Conflict Resolution

In general, the integrity and consistency of information and knowledge must be maintained as design progresses, although local and temporary inconsistency is both allowable and necessary to enable design iteration. Integrity implies the maintenance of functionally related information and knowledge. Consistency is the maintenance of the equivalence of redundant information and knowledge.

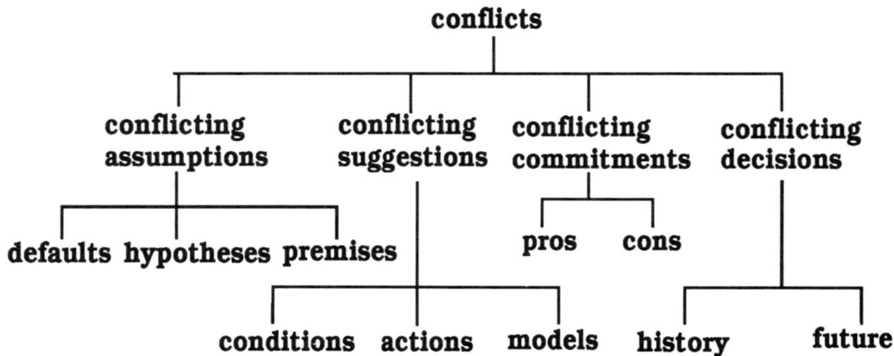

Fig. 4.8 Sources of conflicts

Figure 4.8 shows some possible sources of conflicts. They are grouped according to the classes of design statements:

● Assumptions of two agents are in conflict.
  – *Conflicting default assumptions:* Default assumptions may not be appropriate for many specific situations.
  – *Conflicting hypotheses:* Hypothetical assumptions which conflict are used separately by two agents to draw inferences.
  – *Conflicting premises:* Two agents hold facts which are inconsistent.
● Two agents independently make conflicting suggestions.
  – A conflicting suggestion from two agents has conflicting conditions.
  – A conflicting suggestion from two agents produces a conflicting

decision.

  – Two agents make individual suggestions based on different models which are inconsistent.
- One agent makes a design commitment that leads to an objection by another agent.
  – One agent makes a design commitment one of whose pros is in the set of the cons set of another.
  – One agent makes a design commitment one of whose consequential actions is to add an element into the cons set of another.
- Conflicts exist due to previous decisions.
  – One agent makes a decision which is inconsistent with a previous decision made by another agent.
  – Two decisions lead to further conflicts.

When there is a conflict, it means that there is at least one decision made previously which originated this particular conflict. The strategy for resolving a conflict is to use domain knowledge first, if available, and then try to backtrack to the decision point which contributed to the conflict. Failure handling used in AIR-CYL *(Brown 1984)* is prescriptive. That is, for any action, there can be attached one or more strategies to perform if this particular action fails. Domain-dependent and problem-specific strategies can be encoded by using this type of prescriptive strategy. If this does not work, then it will be necessary to backtrack. *Klein and Lu (1989)* discuss strategies for conflict resolution in cooperative design problem solving at some length. Backtracking strategies are often used recursively to:

- decide when and where backtracking is needed
- explain why it is used
- how it is performed.

The cost in time and design resources is often very high, rising with the extent of backtracking, as a design is progressively decomposed into sub-problems for parallel processing.

Backtracking strategies commonly used include:

(1) ***Chronological Backtracking:*** This strategy reverts to the last decision point if any failure is found at the current point. It is simple and systematic but inefficient.

(2) ***Dependency-Directed Backtracking:*** This strategy intelligently goes back to the last decision point which has contributed to the failure or contradiction *(Stallman et al 1983)*. This is carried out according to records which are created during the constrained reasoning to remember the sources upon which the inference depended. Once an inference source has been found to be contradictory, all dependent inferences and decisions are retracted.

4.4.4 Strategies for Control

Design problem solving involves exploration and backtracking in order to propose plans for solutions, to refine the proposed solutions, and to remedy failed proposals. Both explorations and backtracking are controlled and guided by a knowledge module called the control mechanism. Control of cooperation allows agents to make meaningful contributions whilst maintaining a high rate of convergence towards a satisfactory solution. A control mechanism may use both domain-independent and domain-dependent, and hence problem-specific, knowledge to develop its control rules and strategies.

There are two general strategies for organising control:

- ***Master-Slave (Hierarchical Control):*** Control over a distributed problem solving system may be hierarchical: involving both global and local control, because the system is organised hierarchically. High level problem solvers always have the power to control low level solvers. Control is direct if the hierarchical relations between two solvers are direct.

- ***Same-Class-Citizens:*** Another kind of control regime is based on the presumption that all the experts in the team are equally important. Agents cooperate in the sense that no single member of the group has sufficient information to solve the entire problem; mutual sharing of information is necessary to allow the group as a whole to produce an answer; no one is in total control of the others, although one agent may be ultimately responsible for communicating the solution of the top-level problem to a customer outside the group. In such a situation, individual experts may spend most of their time working alone, concentrating on tasks that have been partitioned from the main task and assigned to individuals.

Sub-problems are said to be nearly independent when they can be solved with little, but not negligible, cooperation in the solution process. Typically, these sub-problems are locally under-constrained, that is, they permit more than one solution when only local constraints are considered. Some sub-problems may be solved separately, but difficulties appear when the sub-solutions are combined. By introducing constraints, interactions and conflicts can be dealt with in terms of the following three types of principles of commitments *(Tong 1987)*:

(1) **The least commitment principle:** The basic idea in the least commitment principle is that decision should not be made arbitrarily or prematurely but postponed until there is enough information available. The least commitment principle states that a decision must not be made until, and unless, all the relevant information is available or no more can be obtained. It is useful when:
   - not enough information is available to ascertain various variables of the sub-systems.
   - the solution to one sub-system may depend on the decisions (or variable bindings) made in the solution of another sub-problem.

By implication, the separation of functions from their least-committed implementations enables the generation of possible solution variants and allows consideration of many alternatives without overlooking relevant solutions.

(2) ***The early commitment principle:*** In contrast to the principle of least commitment, the early commitment principle implies a control strategy where decisions should be made as early as possible. As a consequence, problem solving may involve excessive amounts of conflict resolution, and problems become over-specified. On the other hand by the early commitment principle, the design is increasingly specified structurally as early as possible in the design process. Progressively more detailed information enables the application of a wider range of both qualitative and quantitative analytical techniques. Thus more accurate estimates of system performance can be obtained.

(3) ***The principle of opportunistic commitment:*** This principle provides a compromise strategy for controlling the design process. The key idea is that decisions are made as required and if possible. It moves the attention opportunistically between sub-problems and avoid over-specifying local decisions. In modular design methodologies, functions and their implementations can be mixed based on the opportunistic commitment principle, which suggests that, in a problem domain, some functions be always implemented by one or several fixed modular structures, which are regarded as proven according to current technology.

4.5 SUMMARY

In this chapter, an integrative framework has been proposed for the construction of cooperating knowledge-based expert systems in manufacturing design. A number of significant issues concerning cooperative design problem solving have been discussed. A computational model for cooperative problem solving has been proposed on the basis of agent and meeting metaphors. The model is sufficiently flexible to be capable of incorporating a number of principles found in human design organisations. This chapter has provided an appropriate theoretical model for a practical implementation which is to be discussed in the succeeding two chapters.

Chapter 5
AGENTS: An Object-Oriented Prolog

5.1 INTRODUCTION

Prolog, with its small set of mechanisms, has proven its power in building knowledge-based expert systems for engineering problem solving. The question as to whether Prolog is appropriate for this purpose has effectively been answered by the large number of existing implemented systems. It still remains to be demonstrated whether Prolog will be considered adequate in the long term. *Borrow (1985)* has emphasised the enrichment of computational environments by combining individual tools and constructs effectively to select an appropriate 'cocktail' for each problem. *Subrahmanyam (1985)* has concluded that while Prolog has significant assets along several dimensions, as it exists today, it needs to be modified and enhanced appropriately to make it competitive with extant Lisp systems.

This chapter, based on an initial proposal described in *(Huang and Brandon 1988a)*, proposes an approach to the novel combination of constructs from object-oriented systems represented by Smalltalk *(Goldberg and Robson 1983)* and deductive systems represented by Prolog *(Clocksin and Mellish 1981, and Minker 1989)*. The resulting system, which is called AGENTS, is an Object-Oriented Prolog. It is fully object-oriented, including features such as message passing, inheritance, attributes and constraints (demons). It is also Prolog compatible, so that existing Prolog utilities can be retained intact. Emphasis is placed here on the design issues experienced during experimental implementation of the AGENTS system.

In section 2, the basic ideas of deductive systems and object-oriented systems are analysed in terms of their strengths and weaknesses, and possible combinations are discussed and compared. Section 3 presents two new data structures, agents and objects, which are introduced to enhance Prolog. Section 4 discusses various abstractions built in the AGENTS system. Section 5 demonstrates the key issues related to messages and message passing. In section 6, the implementation of constraints is discussed. The AGENTS system is currently implemented in POPLOG using its Core Prolog utilities. Details on operations of AGENTS and the POPLOG Core Prolog are given in Appendices A, B, C, and D.

5.2 OBJECT-ORIENTED PROLOG

5.2.1 Deductive Systems and Prolog[1]

A deductive system can be considered to consist of a set of clauses with the following form *(Minker 1989)*:

$$P_0 \wedge P_1 \wedge P_2 \wedge \ldots \wedge P_k \Rightarrow R$$

where

(1) P_i – represents a condition.

(2) R – represents the conclusion.

(3) $\Rightarrow$ – is the implication operator.

(4) $\wedge$ – is the conjunction operator.

(5) k – indicates the number of conditions, $k \geq 0$.

> According to the different values of k, clauses can be classified as:
>
> ● facts if $k = 0$;
> ● rules if $k \geq 1$;

---

[1] Prolog is not and does not claim to be the definitive logic programming language. It is not a deductive data/knowledge base either. However, it is the prototypical representative of logic programming languages and is the most widely implemented. The term Prolog is used in this research in the following three senses:

(1) To represent the language Prolog as it exists.
(2) To represent a class of logic programming languages.
(3) To represent a class of deductive data/knowledge bases.

Prolog is a programming language designed in the above fashion *(Clocksin and Mellish 1981, and Bratko 1986)* with additional inference strategies. A Prolog program consists of a set of clauses which are divided into a set of facts and a set of rules. Several of Prolog's advantages can be listed as follows:

(1) **Declarative:** A Prolog program has a primarily declarative syntax. The clauses can be read as implications, universally qualified by the variables occurring in the clauses. The declarative feature of Prolog facilitates declarative statements of domain knowledge.

(2) **Rule-based backward chaining:** Prolog provides a built-in rule-based knowledge representation and backward chained inference strategy.

(3) **Constrained exploration:** The process of execution of Prolog programs can be considered as a kind of constrained exploration. Earlier terms in a clause act as guidelines towards an ultimate goal: if succeeded then explore. On the other hand, subsequent terms are constraints imposed on earlier terms: if violated then backtracking takes place.

(4) **Automatic backtracking:** This is the way that Prolog tests alternatives to find a solution automatically.

(5) **Deduction:** The semantics of Prolog programs are consistent with traditional logic representations. This enables programs to be manipulated in a formal fashion. New facts can be deduced from rules and facts explicitly expressed in the system.

(6) **Pattern unification/matching:** Pattern unification provides the basic computational model for the execution of Prolog programs. A goal is evaluated by attempting to unify it with the head of a rule, and if the unification is successful, then attempting to satisfy the sub-goals in the body. A statement is true if and only if it is logically implied by rules and/or facts in the program.

Prolog has generally been considered as a broadly satisfactory language for dealing with situations involving objects and relations.

The declarative syntax of Prolog programs,   in conjunction with their procedural operational semantics, allows for a combination of declarative and procedural representation of knowledge in a cohesive logical framework.

5.2.2 Object-Oriented Systems

The object-oriented paradigm *(Byte 1986, and Stefik and Bobrow 1986)* is receiving increasing attention from the engineering community *(Barbucean 1984)*. An object-oriented system can be considered to consist of a set of objects. Some of the basic object-oriented constructs are summarised as follows:

(1) **Objects:** "Object" is a very general term. It is commonly used very loosely to refer to any data objects such as classes or instances or to represent real-world entities (domain objects) such as gears and shafts, at an appropriate level of granularity of abstraction for the task in hand. Here, objects are a special data model. The term "instances" is also used interchangeably.

(2) **Classes:** A class is used to represent a set (type, class) of objects which have some similarities in physical, functional, structural, or any other abstract perspectives. Here, agents are used instead of classes.

(3) **Instances:** An instance is used to represent a particular individual entity (as opposed to a class of entities). Here, objects are used instead of instances.

(4) **Methods:** This is what an object uses to construct a response to a particular message. Methods are procedures held locally by a class of objects and are available to all instances of that class. The execution of methods brings about changes in the state of an object.

(5) **Message Passing:** This is how objects communicate with each other and how the outside world communicates with an object. Message passing changes the conventional way of applying a procedure to a data object to that of a data object deciding what and how to do according to receiving messages. Here, messages are ordinary Prolog query clauses.

(6) **Attributes:** Attributes describe the static properties of an object. Values of attributes represent the state of an object. Instances of the same class have the same attributes but can have different values for them.

(7) **Inheritances:** Inheritances define ways of sharing information. That is, when a class of objects is specified as having characteristics in common with an existing class, but with some different or extra attributes, methods or characteristics, then it can be said to inherit from the first class.

(8) **Demons or constraints:** Constraints or demons are active procedures to be executed when certain actions affect properties of an object. For example, there are demons for attributes and methods.

The above constructs offer a number of useful features. It is quite often possible to enhance a system using object-oriented constructs, as will be done to enhance Prolog in this chapter. One major advantage is that an object-oriented system offers a high degree of modularity. Each agent appears as a black box to others. It is not necessary for agents to possess knowledge of each other's internal structure and methods.

Another major advantage of object-oriented systems is the capability of rapid prototyping. This characteristic is closely related to that of modularity. That is, the design of the system does not require connections between objects to be carefully planned and defined in advance, with the exception that inheritance relationships between classes have to be specified. The primary benefit of inheritance is that shared information need not to be duplicated.

5.2.3 Pitfalls

Developers of knowledge-based expert systems using Prolog have tended to structure and organise knowledge and information in the form of modules *(Topping and Kumar 1989)*. The central requirement is that they must ensure that there are no duplicate predicates in different modules. Unwanted duplication of predicates might lead to misunderstanding and erroneous performance.

Most engineering problems are characterised by a large number of domain objects with high complexity. Prolog has been found limited in the capability of its data abstraction for this purpose. The basic facilities provided by Prolog for knowledge representation and inference are somewhat sparse, allowing only a flat collection of facts and rules for knowledge representation and backward chained inference. In summary, Prolog suffers from lack of facilities to modularise knowledge, lack of facilities to describe complicated concepts or objects, and lack of facilities to construct hierarchies or networks between them. As a consequence, system developers are forced to establish special structures based on a limited number of Prolog constructs to meet contextual requirements.

Criticisms of object-oriented systems have been currently eclipsed by enthusiasm of their applications: any system which has some object-oriented features is implicitly and uncritically considered as being 'good'. This chapter does not set out to contribute to the discussion of the limitations which an object-oriented system might suffer from, rather it is maintained that object-oriented constructs can be used positively to enhance existing Prolog based systems.

5.2.4 Combination: Logical Objects versus Object-Oriented Logic

Regardless of individual disadvantages and advantages of Prolog and object-oriented systems, it is of significant value to combine the constructs from the two types of systems. It has been found possible for Prolog to provide object-oriented constructs, as has been attempted in DAISIE by *Ishii et al (1989)*. It has also been found desirable for an object-oriented system to contain Prolog-like features *(Tomiyama et al 1989)*. Such a combination should exploit individual advantages without causing any extra problems. To achieve a hybrid performance mixed with object-oriented and logical constructs, there are several approaches, which may be summarised as follows:

(1) ***Symmetrical Hybrid Environments:*** A hybrid environment with both Prolog constructs and object-oriented constructs mixed has been proposed by *Borrow (1985)* and provided within the POPLOG programming environment *(Laventhol 1987)*. In this approach, problems of interface between sub-systems may be intensive.

(2) ***Hybrid Systems:*** A system provides multiple schemes for representation and a variety of inference strategies, as reviewed by *Allen et al (1987)*.

(3) ***Object-Oriented Logical Systems:*** This builds object-oriented constructs on the top of Prolog *(Ishii et al 1989)*. The resulting systems inherit the disadvantages of existing Prolog systems discussed previously.

(4) ***Logical Object-Oriented Systems:*** This is to build logical constructs on the top of object-oriented paradigm *(Tomiyama et al 1986)*.

(5) ***Object-Oriented Prolog:*** This extends the existing Prolog internally to integrate object-oriented constructs with extended built-in predicates while maintaining maximum compatibility with existing Prolog syntax and semantics.

(6) ***Other Approaches:*** There are other proposed methods to integrate logical programming with object-oriented constructs *(Tokoro and Ishikawa 1988)*. These systems have lost their dialect of Prolog or any particular object-oriented systems. Therefore, the acceptance for popular use from the engineering community will be uncertain.

5.2.5 Object-Oriented Prolog

This text develops an Object-Oriented Prolog approach. It aims to construct an Object-Oriented Prolog which provides the advantages of both Prolog and OOP while reducing both their combined and individual disadvantages to a minimum.

The fundamental principle of the approach is that the resulting system must retain the maximum compatibility with existing Prolog syntax and semantics. One of the simple reasons for this is that Prolog has been widely recognised and accepted as a powerful programming language and a substantial reservoir of expertise has already been coded and accumulated in Prolog in many fields of engineering.

Some of the general considerations for extending a Prolog into an Object-Oriented Prolog are summarised as follows:

(1) ***Agent Metaphor:*** In Prolog, the basic data structure is the term. Clauses are constructed from terms. In an Object-Oriented Prolog, a new data structure called agent is provided at a higher level. An OO Prolog program can then be considered as a set of separate agents and each of the individual agents comprises an ordinary Prolog programme consisting of clauses.

(2) ***Clauses as Methods:*** The concept of a method is one of the essential object-oriented constructs. Methods determine what an agent should perform, and how, in response to its interaction with the outside world. User-defined clauses in an agent are considered as methods.

(3) ***Queries as Messages:*** Message passing is the fundamental mechanism for inter-agent communication in object-oriented systems. Agents are distributed among an OO Prolog program, and therefore queries to the system must be directed to specific agents. Targeted queries are taken as messages.

5.3 AGENTS AND OBJECTS

5.3.1 Modularisation

There are two different levels of modularisation, at clause level and at agent level, in the AGENTS system. An AGENTS program consists of agents and, in turn, an agent consists of clauses. Clauses are basic elements of agents, and are both local and private. Therefore, different agents may have clauses defined in the same way (for example, with the same functor and arity).

As far as the modularisation at the agent level is concerned, there are two concepts: classes and instances. Some object-oriented systems provide a uniform representation of classes and instances, whereas some systems make a distinction between them. The latter approach is adopted here: agents represent classes and objects represent instances.

This distinction between classes and instances is useful since an agent can be considered as a generic problem solver, whilst an object can be thought of as a result describing a specific state of the agent. For example, an agent called "gear" can be defined to be responsible for designing gears. The "gear" agent designs all the gears needed in a specific transmission box. Therefore, the process of agent-based problem solving is that of creating objects and instantiating their attributes with values.

5.3.2 The Agent Data Model

Computationally, agents are merely a data structure for manipulation (including storage and retrieval) of information. The agent data model consists of a number of components:

(1) **Agent Name:** Every agent has a name for unique reference in the system. The system does not allow agent duplication. This is avoided in the agent creation procedures: if there is an agent with the same name as being defined, then it is assumed that the existing agent is being updated.

(2) **Inheritance Pointers:** These are used to describe the relationships between agents. They have a great effect on inheritance of properties. This is an essential feature of an object-oriented system.

(3) **Attributes:** These describe the static properties of classes of objects. For example, the design specification of a gear is determined by a number of technical parameters including its module, teeth number, pitch diameter, pressure angle, helical angle, etc. The task of design problem solving is to instantiate the values of attributes under a number of constraints.

(4) **Methods:** Methods in this Object-Oriented Prolog are ordinary Prolog predicates defined by clauses. They are compiled into procedures and retained within an agent that can use them when

appropriate messages are transmitted.

(5) **Demons:** Demons are also represented by special Prolog clauses with special functors. Demon clauses are also compiled into procedures and kept in an agent. Their uses, however, are associated with certain actions of method execution. They can be used as preconditions (before methods) or post-actions (after methods) for a primary method.

Agents can be defined in an OO Prolog program.   An agent definition is bracketed within two special clauses:

```
agent(Name).
    ...
endagent.
```

or

```
agent(Name, Reconsult).
    ...
endagent.
```

where

(1) agent(Name) – This clause starts a definition of an agent with the name of Name.

(2) endagent – This clause ends the agent definition.

(3) Reconsult– This argument is optionally used to indicate whether clauses should be overridden if there is already an agent with the given name Name in the system.

Agents can also be defined dynamically. In this case, however, properties must be created and added in subsequent actions. Following is a clause for defining an agent with the given name Name, and creating an attribute of Attribute for this new agent:

```
create(Name, Attribute) :-
    agent(Name),
    Name ?- attribute(Attribute).
```

There is a special agent with the name "meta" in the AGENTS system. It provides all the built-in predicates. Some of these are similar to standard Prolog built-in predicates; some are designed for managing object-oriented constructs; and some are intended for their integration. Appendix D presents a description of all the built-in AGENTS predicates.

5.3.3 The Object Data Model

Objects are another kind of data structure used in the AGENTS system. These are used only for storing and retrieving information. In comparison with the agent data model, the object data model has the following characteristics:

(1) Agents have all the four basic sets of properties: attributes, pointers, methods and constraints. On the other hand, objects have only two: a set of attributes and a set of values corresponding to attributes. The attribute set of an object includes all the attributes defined by its template agent and those inherited from the super-agents. The values of object attributes are assigned incrementally by the execution of methods of agents.

(2) In fact, objects can be considered as a kind of properties of agents. Objects belong to agents. An agent can have as many objects as is necessary to accomplish its purpose. However, through inheritance, an object implicitly has all the properties that its agent has. Objects of the same agent share their overall characteristics although the values of attributes may differ from each other.

(3) Objects can be used to represent different versions of the corresponding agents during an evolutionary process of problem solving. For example, an initial object can be created when the agent starts to solve a problem. Such an object may have only a few attributes with roughly estimated or even unknown values. After a sequence of methods have been attempted, intermediate objects can be created, as necessary, to record the developing history. When a goal state is reached, the final object is created with the most specific information. The final object represents the result, and all the versions created during the agent development constitute the solution.

In an OO Prolog program,   an object should be defined within the agent definition:

```
agent(student).
    ...
    object(huang).
    ...
endagent.
```

An object can be defined dynamically by sending an appropriate message to the corresponding agent:

```
student ?- object(huang).
```

or equivalently

```
meta ?- object(student, huang).
```

5.4 ABSTRACTIONS AND INHERITANCES

5.4.1 Attributes

Agents are necessarily attributed. In turn, attributes can have facets. Attribution is the basic mechanism in object-oriented representation which derives from related types of representation for data abstraction such as frame-based and semantic-net structures. Attributes represent static agent properties which can be perceived in advance (during the process of complexity management). It is admissible for different agents to have attributes with the same names. For instance, a gear agent has an attribute with the name of diameter while a shaft agent also has an attribute with the same word. They share this similarity because they can be classified into an enveloping category of rotational bodies.

Typing mechanisms have not been incorporated in AGENTS, just as Prolog does not have any typing functions. A requirement of the data models of the AGENTS system is that it is aimed at supporting conceptual design as well as detail design. In the transition between these two cases, the types of attributes of the same agent are likely to vary progressively from symbolic to numeric. A strong typing mechanism would inhibit this process of evolutionary change.

Attributes can be defined either statically, during compilation, or dynamically, at run-time. The built-in predicate ***-attribute-*** is available for declaring a word as being an attribute of an agent. It must be used within the agent definition:

```
agent(gear).
      attribute(name).
      attribute(module).
      attribute(teeth).

      ...
endagent.
```

or as a message passed to an agent:

```
gear ?- attribute(diameter).
```

Attributes of an agent have default values, whilst attributes of an object must have values assigned. The built-in facilities for access and update of these two types of attributes are listed in table 5.1.

Table 5.1 attribute access and update

agent attributes and default values	
default access	**Agent ?- default(Attribute, Default).**
default update	**Agent ?- attribute(Attribute, Default).**
object attributes and values	
value access	**Agent ?- value(Object, Attribute, Value).**
value update	**Agent ?- instantiate(Object, Attribute, Value).**

5.4.2 Methods and Demons

User-defined clauses of an agent are called methods (cf. clauses for defining attributes and inheritance pointers). There are two kinds of methods:

```
Head.                    – facts
Head :- Body.            – rules.
```

Traditionally, methods are used to represent capabilities, rather than to store data in object-oriented systems. However, in AGENTS, methods can be used to store data in the same way as any standard Prolog clauses.

Demons (constraints) are active procedures to be executed when certain actions are taken, associated with defined method clauses. They are attached to appropriate predicates by using a special syntax

```
Head := Body.      – for "after" constraints
Head =: Body.      – for "before" constraints
```

In an AGENTS program,   methods and constraints can be defined in exactly the same way that Prolog clauses are defined, as long as they are bracketed within agent declarations. At run-time, methods and constraints can also be defined dynamically or cancelled using the built-in clause manipulation facilities: "assert" and "retract" in the same way that clauses can be asserted to or retracted from an ordinary Prolog program.   However, care should be taken to ensure that an assertion or a retraction is made within the right agent. There is little problem when a message-passing mechanism is used:

```
Agent ?- assert(Clause).      – add Clause to Agent
Agent ?- retract(Clause).     – delete Clause from Agent
```

A method, either primary (a fact or a rule) or a constraint, is defined declaratively by a predicate consisting of a set of clauses with the same functor and arity. After all, a method is a procedure which will be executed during the process of message passing. There are two major parts of a method procedure:

- pattern matching / pattern unification
- tail resolution / body evaluation

A method procedure is successfully executed if and only if patterns are matched and the conditional body is resolved. Appendix B discusses general aspects of pattern matching and body evaluation.

Special methods are built-in predicates defined in the meta agent and inherited by all the other agents.

(1) *willrespondto(Agent, Goal):* This predicate tries to check if there is a method defined in this Agent corresponding to the nominated Goal. It succeeds if and only if there is a method defined in the receiver; otherwise it fails.

(2) *dontcare(Agent, Goal):* This method checks if there is a method defined in this Agent corresponding to the nominated Goal. It fails if there is such a method but the method itself fails. It succeeds if there is no corresponding method or the method which is found itself succeeds. The predicate *dontcare(Agent, Goal)* can be defined using -*willrespondto*- method:

```
dontcare(Agent, Goal) :-
        willrespondto(Agent, Goal),
        Agent ?- Goal.
dontcare(Agent, Goal).
```

(3) anymessage methods: Sometimes a special action must be performed when any message is passed to an agent. In this case, it would be very tiresome to have to write a method for all the possible messages. Instead, a special syntax is provided to define a method, without a pattern head, for any messages:

```
:- Body.
```

5.4.3 Inheritance Pointers

A feature common to almost all object-oriented systems is that of inheritance. This allows the user to define one class (agent), such as "person", which encapsulates all the relevant features of human beings, and then to define another class (agent) which belongs to the same class but is a specialisation of it. For example, a "student" is a kind of "person" and has all the attributes and characteristics that human beings have but might also be considered to have some special attributes or characteristics. A "lecturer" is also a kind of "person" who has some special attributes and characteristics. A "student" is different from a "lecturer" in that students attend lectures and lecturers present lectures though they will share attributes such as "subject".

The inheritance mechanism provides two facilities for indicating relationships: "inherit from" and "inherit by". The "student" agent "inherits from" the "person" agent and on the other hand the "person" agent is "inherited by" the "student" agent.

Inheritance pointers can be specified, within the agent declaration, by two built-in predicates:

    super(Agents)
    sub(Agents).

    Following are three example agents:

        agent(person).
            sub(student).
            sub(lecturer).
        endagent.
        agent(student).
            super(person).
        endagent.
        agent(lecturer).
            super(person).
        endagent.

In general, inheritance pointers are required to enable classification. The sub agent is a specialised form of the super agent. Therefore, the sub agent inherits all properties possessed by its super agent. Inheritance pointers are also used for projection: i.e. viewing an entity from a number of perspectives to obtain a variety of images. (See Chapter 3 for fuller discussion of the projection abstraction). The primary entity and its projected images can all be represented by agents. An image agent deals with a restricted range of aspects of the corresponding entity agent. The union of the set of image properties comprises all of the aspects of the entity agent. To accomplish this, images can be represented by super agents and the entity by a sub agent inheriting from all of its constituent image agents.

5.4.4 Property inheritances

Inheritance management uses the structural links of an abstraction network to augment and propagate both descriptive and behavioural properties. In the AGENTS system, there are two types of basic abstraction: classification and instantiation. Accordingly, there are two types of inheritance: class inheritance and instance inheritance.

(1) **Class Inheritance:** Class inheritance specifies the sharing of information between agents at different levels. Attributes and methods of an agent are inheritable from a super class of agents to sub-classes of agents. An individual agent can inherit from more than one super agent. Instead of copying properties from supers to subs, inheritances are made dynamic rather than static. This may retard the processing effort but compensates by facilitating property management. For instance, a property may be cancelled during running time; any reference to the property in any later actions would then be made invalid unless the sub agent has a property defined with the same name. Class inheritance is achieved by the message passing and processing mechanisms.

(2) **Instance Inheritance:** Instance inheritance specifies sharing of information between an agent and its objects. Instance objects inherit all the properties from their template agents. The instance inheritance is automatically and implicitly made within the Object-Oriented Prolog system. However the end users may not necessarily be aware of this. Instantiation inheritance is achieved at the time that a specific object is created from a template agent. The message-passing mechanism does not deal explicitly with instantiation inheritance since messages are passed to agents (classes) instead of objects (instances) in the AGENTS system. This differs from most other object-oriented

systems where messages are sent to instances instead of classes.

5.4.5 Precedence List (Inheritance Lattice)

As has been mentioned previously, in standard Prolog it is important to take great care that multiple definitions of variables cannot occur. In AGENTS these requirements are relaxed at a cost of the provision of precedence rules wherever consequential ambiguities occur.

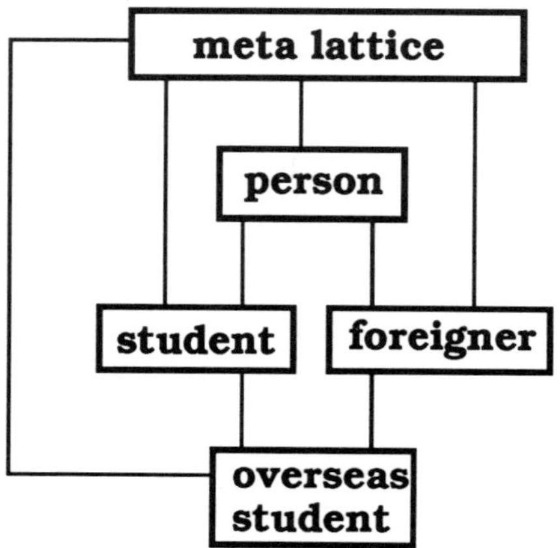

[os-student,student,person,foreigner,meta]

Fig. 5.1 construction of inheritance lattice

Consider, for example the inheritance lattice shown in figure 5.1. The "overseas student" agent inherits from both the "student" agent and the "foreigner" agent. Both the "student" agent and the "foreigner" agent inherit from the "person" agent. All these agents inherit from the meta agent by default. Suppose that there is a method concerning 'language' defined in the "student" agent. There is also a method concerning the same term 'language' defined in the "foreigner" agent but with a different implication. The student agent must understand the language in which lectures are given. The foreigner agent may "speak" a foreign language. What happens if a message concerning 'language' is sent to the 'overseas student' agent? Obviously there is now an ambiguity whether to use a common lecturing language or a foreign language? When such a conflict arises, the AGENTS system uses a precedence list to determine intended action. The way of building up a precedence list is complicated by the capability to use multiple inheritances and abstractions. The general rules for constructing a precedence list is as follows:

(1) The agent itself is the first in the precedence list:

p-list = [overseas_student|_]

(2) Then, the precedence list of the first super agent of the agent is linked with previous p-list:

p-list = [overseas_student, student. person|_]

(3) Repeat (2) over all the super agents:

p-list = [overseas_student, student. person, foreigner, person]

(4) There should not be any repetition of agents:

p-list = [overseas_student, student. person, foreigner]

(5) The meta agent is always the last element of the precedence list to ensure that every agent inherits from the meta agent:

p-list = [overseas_student, student. person, foreigner, meta]

5.5 MESSAGE PASSING AND PROCESSING

5.5.1 Communication by Message Passing

In ordinary Prolog programs, queries are clauses that constitute bodies but with empty heads, and are issued interactively from the user. Users are able to ask a Prolog program what is true:

```
?- grandfather(smith, Who).
```

This question will be answered by a rule:

```
grandfather(X, Z) :-
        father(X, Y),
        father(Y, Z).
```

This rule will issue another two queries to the system to ask who is the father of 'smith' and who is the father of Smith's father, as shown in figure 5.2 (a). Prolog accepts the following three types of queries:

(1) a single query;
(2) a list of conjunctive queries; and
(3) a list of disjunctive queries.

A list of conjunctive goals is true if and only if all of the individual queries are true for the same assignment of variables. A list of disjunctive goals is true if and only if at least one element in the list is true.

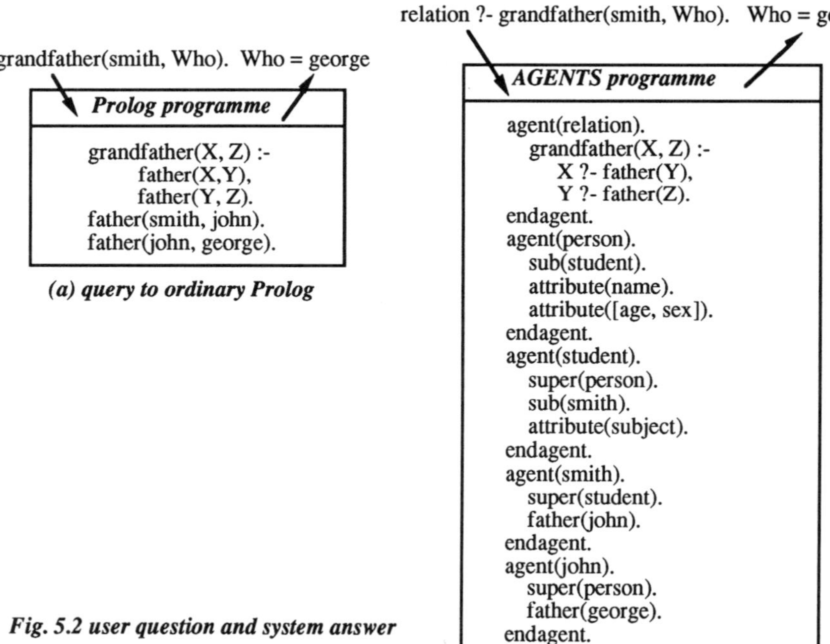

(a) query to ordinary Prolog

Fig. 5.2 user question and system answer

(b) message to AGENTS

In Object-Oriented Prolog, in contrast, queries must be directed to agents to ask what assertions are true (or at least believed). Figure 5.2 (b) shows how the same query as used in figure 5.2a is processed by a simple AGENTS program with a query made to the "relation" agent. There is a rule for defining the relationship of "grandfather" in the family "relation" agent. When the user question "Who is the grandfather of Smith" is asked to the "relation" agent, the "grandfather" method sends a message to "smith" to ask who is the father and send another message to Smith's father to ask who is the father.

5.5.2 Message Passing and Interpretation

Message passing is a central mechanism for communication in object-oriented systems. Messages are the only medium for agents to communicate with each other. Agents sense their environment and determine their course of action through messages.

The user must indicate to which agent a question is addressed in an OO Prolog program. Such a query with a target is called a message. A message, in general, is represented as a special Prolog term of the following form:

Receiver ?- Goal.

where

(1) Receiver – It is the agent to which the question is asked.
(2) Goal – The question either from the user interactively or from other messages.
(3) ?- – This is the principal functor of the message

The information which indicates the sender who issues a message, the time that a message is issued, etc. is not included in the message format. It is sufficient to select an appropriate method based on a message with the above two essential arguments: receiver and query. Messages are passed to agents rather than objects in the AGENTS system, i.e. the receiver must be an agent not an object.

In a multi-agent environment, there are two special agents representing the state of message passing: the current agent and the self agent. Their roles will be described subsequently. The receiver of a message may sometimes be omitted. However, this is not allowed in message processing. A procedure called a message interpreter is used to resolve this situation, as shown in figure 5.3. The output from the interpreter is a message. The input to the interpreter can be one of the following:

(1) An ordinary message Receiver ?- Goal. – In this case, nothing needs to be done.

(2) A non-receiver built-in predicate – The meta agent becomes the receiver.

(3) A self message ^Goal. – In this case, the "self" agent is the receiver.

(4) An ordinary goal Goal – In this case, the current agent is assigned as the receiver.

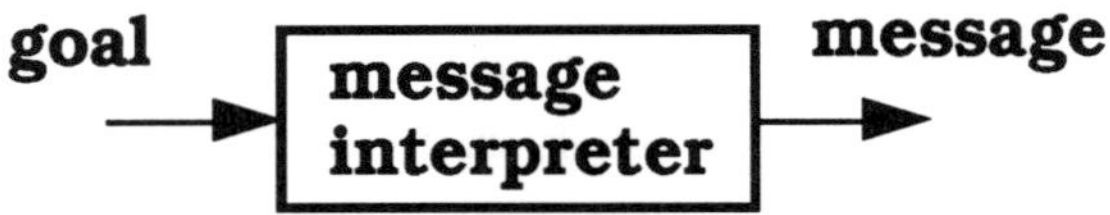

Fig. 5.3 *message interpreter*

Figure 5.4 shows an agent in which a clause is defined with three terms in its body. The first term $Term_1$ is a message with the receiver *Receiver*. The second term $Term_2$ is also a message but its receiver must be an *Agent* which will be remembered as a particular agent - the "self" agent. The third term $Term_3$ does not have a receiver explicitly but the current agent will be assigned as its receiver.

```
agent(Agent).
        Head :- Receiver ?- Term₁,
                ^Term₂,
                Term₃.
endagent.
Agent ?- Agent ?- Head.
```

Fig. 5.4 *example of sub-goals*

An agent may be invited to perform a sequence of actions continuously within its own context. The invitation mechanism is designed for this purpose: all the queries without specified targets (receivers) are defaulted direct to the invited (current) agent. Once an agent is invited for a session, the following actions take place within its private context until changed again by a subsequent invitation to another agent.

The invitation mechanism provides a way of re-using knowledge repeatedly. An agent has a rule segment where the bodies are not sufficiently complete for them to become messages (in that they have a goal but the receiver is not yet nominated). When such a rule is activated by external messages at different times, it will perform differently. The reason is that the current agent changes as design progresses and the nature of the appropriate response varies. In consequence, those terms which are not yet messages will be assigned different receivers corresponding to current agents.

For instance, suppose that the system is currently active within the user agent, as prompted by predecessor of the "?-" symbol, when an invitation is sent to the gear agent. An interactive process can be demonstrated as follows:

```
user   ?- invite(gear).                    ;;; invite the gear agent for a session
       => yes.
gear   ?- design.                          ;;; issue the command to design
       => yes.
gear   ?- transmission ?- completed(gear). ;;; send a message to transmission agent
       => yes.
gear   ?- invite(user).                    ;;; return to the default agent (here user)
       => yes.
user   ?- _                                ;;; ready for further actions
```

Invitation does not guarantee that sub-goals be executed within the same context of an agent since sub-goals themselves may change the current agent (by issuing an invitation to a new agent). Consider the example in figure 5.4. In this case, the receiver of $Term_3$ is not specified. As a consequence, the current agent will be used as the receiver by default. Suppose that Agent is the current agent at the time when the following message is issued:

Agent ?- Head.

After $Term_1$ and $Term_2$ are processed, the current agent may be changed to some other agent (if an invitation is issued when $Term_1$ or $Term_2$ is processed). Therefore, $Term_3$ will be processed within the context of the new current agent.

It is also possible to ask questions to the agent that is currently asking a question. This special agent is kept as the "self" agent. The following special syntax is provided for representing the case of sending a message to the "self" agent:

^Term.

which is equivalent to

self ?- Term.

In the example of figure 5.4 $Term_2$ is a message sent to the "self" agent, Agent in this case, and is processed within the context of the Agent agent.

In general, the target argument of a message must be an atom which denotes the target agent. The case with a variable target is complicated, which requires a different type of protocol of communication – broadcasting (see Chapter 3 for a brief discussion). A very simple broadcasting strategy may be 'passing a message to an agent sequentially in the order that they were created'.

5.5.3 Message Processing

There are three procedures for processing messages. They include:

(1) **Core Message Processor:** The key task of the core message processor is to run a clause procedure over a goal. The procedure introduces the arguments of the goal as local arguments. The way that the core message processor operates is affected by any constraints which are introduced (see below for discussion on constrained message processing).

(2) **Message Handler:** When an agent receives a message, an appropriate method is selected and the associated procedure is executed. Agents are black boxes: passing a message or running a method is a private process. Agents understand messages from the outside world and make what-to-do and how-to-do decisions on their own behalf. Each agent has a specific message

handler for this purpose. This is a partly compiled procedure with preparatory processing being done before it is actually executed. For instance, the precedence list of the agent, and all the methods and constraints, including those inherited from its super agents, are prepared in advance. Thus, the message handler manages inheritance of methods and constraints. The primary actual task of the message handler is to call the core message processor to complete the whole job.

(3) **Message Processor:** This is the procedure used for actually processing messages. It utilises the message interpreter and the message handler to accomplish its task.

5.6 CONSTRAINTS ON PREDICATES

5.6.1 Types of Constraints

One of the features of object-oriented systems is the provision of a way of defining constraints (also called demons). These are active procedures, to be executed when certain actions take place, with associated properties such as attributes or methods or even objects.

Some typical demons attached to attributes are:

- ***if-changed:*** The demon is fired when the value of its attached attribute is changed (updated).
- ***if-needed:*** The demon is executed when the value of the attribute is needed (accessed).
- ***if-removed:*** The demon is fired if the attribute is deleted from the agent.
- ***if-added:*** The demon is fired if the named new attribute is defined for the agent.

Some of the typical demons attached to methods include:

- ***if-selected:*** The demon is fired if (before) the primary method is executed. This can be treated as precondition for the primary action.
- ***if-succeeded:*** The demon is fired if (after) the primary action succeeds. This can be considered as the post-action.
- ***if-failed:*** The demon is fired when a goal fails. This can be viewed

as action for failure recovery.

There may also be some demons which are active at the level of agents and objects, perhaps modifying the control of message passing. For instance, any-message methods. The '-before- any-message' can be viewed as an entry constraint on inviting an agent for contribution. The primary 'any-message' method again is executed before any other methods. The '-after-any-message' method is an exit condition from an agent. (See above for discussion on any-message method).

5.6.2 Demon Implementation

Features of this type are extremely useful in cooperative design problem solving to explore new goals or to resolve conflicts cooperatively. Currently, three types of demons are included in the agent implementation:

- before (=:),
- after (:=), and
- if_failed (-:).

The following is an example of an agent with demons:

```
agent(Agent).
        attributes(Attributes).
        ...
        Functor(Arguments) =: Before1.
        Functor(Arguments) =: Before2.
        Functor(Arguments) :- Primary1.
        Functor(Arguments) :- Primary2.
        Functor(Arguments) := After1.
        Functor(Arguments) := After2.
        Functor(Arguments) -: If_Fail1.
        Functor(Arguments) -: If_Fail2.
        ...
endagent.
```

There can be any number of constraints on one predicate as

there can be any number of clauses for a predicate. In the above example, there are two clauses for the primary predicate method. There are two 'before' and two 'after' clauses attached to the primary predicate. Two if-failed demons are also included. They share the same functor and the same number of arguments (arity). However, they may differ in the patterns required for triggering their execution. The first matched is the first to be tested and the first which succeeds is the solution.

When a query message is sent to the agent:

Agent ?- Functor(Arguments),

the message is processed under constraints following the flow chart shown in figure 5.5. There are two kinds of activities:

(1) *Checking the existence of an appropriate clause:* The existence of a primary method is essential to the result.

(2) *Execution of the corresponding clause procedures:* Any failure in execution of clause procedures results in an execution of an if-failed demon clause, if there is one, or the final failure to backtrack.

The goal fails ultimately if one of the following occurs:

(1) There is not a primary method in the receiving agent for the message.

(2) The "before" demon fails.

(3) The primary method or "after" demon fails, and "if-fail" demon fails.

The goal succeeds if one of the following occurs:

(1) Corresponding methods including "before" and "after" demons execute successfully.

(2) The primary method or the "after" method fails but the corresponding "if-failed" demon succeeds.

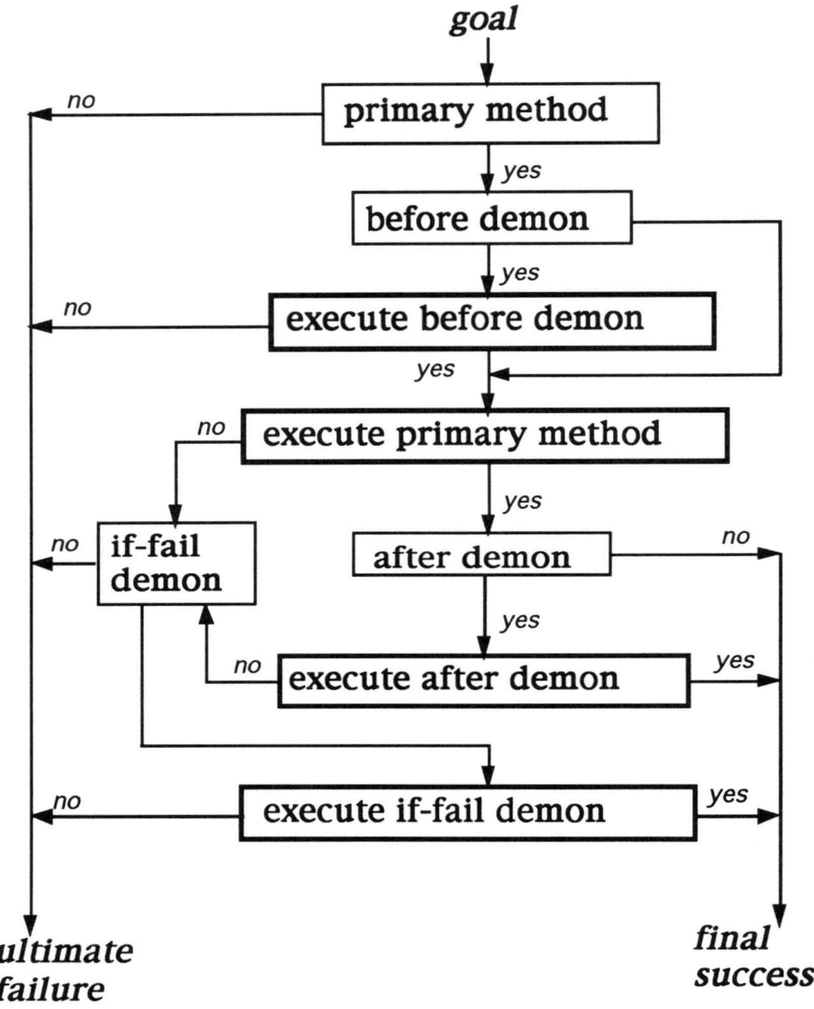

Fig. 5.5 constrained message processing

5.6.3 Issues with Demons

Several issues must be discussed concerning the improvement of
the implementation of demons:

- **Demon inheritance:** Demons are active properties which, as a
  consequence, are inheritable. Whether or not demons should be
  inheritable, and how they should be inherited, are open to

discussion. Here for the simplicity of experimental implementation, demon inheritance is treated in the same way as method inheritance.

- **Pattern Variables:** The execution of demons may partially instantiate pattern variables of the primary Prolog clause. For instance, the execution of the "before" demon associated with the primary message will instantiate some of the variables. This will result in a change in the pattern and further affect the matching of primary clauses.

- **Control:** Demons potentially affect backtracking. If any "before" demon fails, the primary goal fails; if any "after" demon fails, the primary goal also fails. If there is no demon defined or there is no demon which matches the pattern, there should be no action, the default being 'successful'.

- **Declarative:** Prolog is already a constrained programming language viewing conditions of rules as constraints on goals. The introduction of explicit constraints on predicates provides a declarative style of constraint definition: constraints can be defined or deleted freely without affecting the primary predicates.

5.7 SUMMARY

A number of issues have been discussed concerning the extension of Prolog into an Object-Oriented Prolog, including the following main points:

- the agent data model
- abstraction and inheritance
- message passing
- constraints (demons)

With the introduction of agents and objects, the central Prolog data base is divided into a number of inter-dependent deductive knowledge bases. This is characteristically desirable for problem solving in fields such as design, planning, etc. The resulting Object-Oriented Prolog has extended object-oriented features without sacrificing standard Prolog features. This will prove attractive to the majority of Prolog users.

Chapter 6
Programming in AGENTS

6.1 INTRODUCTION

Prolog has been widely accepted as a powerful programming language for the development of knowledge-based systems in a number of engineering fields, for instance, architectural design *(Roseman et al 1985)*, structural design *(Topping and Kumar 1989)*, and manufacturing design *(Kim 1985)*. In addition, Prolog is not limited to the building of knowledge-based systems. It is a general-purpose programming language as well. In this particular context, however, Prolog is considered as a tool only for knowledge-based system development. Moreover, Prolog itself could be viewed as a type of primitive expert system shell with basic components: the working memory, the knowledge base, and the inference engine. Other functional components can be designed with ease.

In the preceding chapter, a number of methods have been devised to integrate constructs from both Prolog and Object-Oriented systems. This extends standard Prolog to provide a more suitable tool for developing cooperative knowledge-based systems, primarily for design problem solving. The resulting Object-Oriented Prolog is substantively similar to ordinary Prolog wherever equivalent constructs are compared *(Huang and Brandon 1992)*. Detailed discussion of the operation of Prolog is outside the scope of this work, since it is covered adequately in a number of standard texts, for example *Clocksin and Mellish (1981)* and *Bratko (1987)*.

The discussion here is centred on a number of facets of expert systems, including knowledge representation, inference strategies, and user interaction and decision explanation. It should also be noted that many of the features presented are currently experimental or under development.

This chapter discusses the usage of the AGENTS system and explains how the extended Object-Oriented Prolog can be used to develop and coordinate knowledge-based systems. Section 2 summarises the general features of the AGENTS system. In section 3, the architecture of the AGENTS system for cooperating knowledge-based systems is discussed with reference to the blackboard metaphor. Section 4 presents the architecture of individual agents, emphasising the aspects of user modelling. Section 5 discusses issues of knowledge-based development using AGENTS, in comparison to ordinary Prolog development.

6.2 AGENTS FEATURES

6.2.1 General Features

The AGENTS system is the condensation of the theories developed over previous chapters. It achieves a novel combination of constructs from both deductive and object-oriented systems. It is designed to be compatible with Prolog, in terms of semantics and syntax, since many of the existing knowledge-based systems in manufacturing design have been implemented in Prolog. Existing Prolog code can be used within individual agents without any change. In addition, the AGENTS system is equipped with some of the knowledge-based facilities such as an explanation capability and a blackboard structure. In summary:

- It integrates declarative and procedural features.
- It provides controlled access to shared data through inheritance.
- It offers high modularity in structuring knowledge.
- It achieves integration of data and procedures.
- It enables effective methods of data abstraction.

6.2.2 AGENTS Programs

There are two ways of viewing an AGENTS program:

(1) It is composed of a set of agents, each of which comprises up to five optional parts in its declarations: attributes, pointers, methods, constraints and objects.

(2) It consists of a set of clauses which can be divided into six categories: agent declaration, attribute declaration, pointer declaration, method declaration, constraint declaration, object declaration.

Property declaration must be bracketed within the agent declaration. The order in which agents are defined is trivial and of no significance. However, the order of clauses within agent declarations is significant. For instance, the order in which inheritance pointers are defined affects the process of building the inheritance lattice, and the order in which methods are defined determines the sequence of clause execution.

6.2.3 AGENTS Utilities

The primary role of the AGENTS system is for the development of Cooperating knowledge-based Systems. There are at least two styles to use AGENTS for cooperative problem solving:

(1) Every agent in the community is a problem solver, enabling the community as a whole to solve a number of problems. What is intended is that the community can solve problems which are dependent on, but larger than, the problems which are within the capability of individual agents. Such a collective capability relies on cooperation between agents. The user describes a problem to an agent in the AGENTS system. This appointed agent is then responsible for finding an appropriate method for solving the user's problem. The problem may be solved within the context of this single agent but, more often, the selected agent, through its method constructs, will consult other agents for assistance by issuing task messages to them. A consultant agent treats a task defined in terms of a message as a problem

and attempts to generate a solution. In such a fashion, a solution to the original user problem is generated.

(2) There is one agent which acts as the central problem solver for making inferences and decisions. The remaining agents, excluding the "meta" agent, in the community are considered as distributed knowledge bases. The user describes a problem to the agent which acts in the role of central problem solver. An appropriate method is then found for solving the problem. The selected method uses information and knowledge stored in other agents. Retrieved knowledge refers to new knowledge belonging to different agents. Relevant knowledge is searched and tested for appropriateness in the defined application. In this way, a successful solution is generated.

The secondary objective of the AGENTS system is to provide a robust data structure with a rich set of data abstractions for modelling and model integration:

(1) Attributes can be used to describe static and structured data.

(2) Methods retain the flexibility of modelling activity and behaviour which make use of problem description.

(3) Although the AGENTS system does not provide explicit facilities for multi-view, multi-version, multi-context, and multi-component modelling, it retains full flexibility for users to establish their own models for specific problems.

(4) The AGENTS system achieves the integration of problem description with problem-solving activities.

(5) In addition, data transfer is made possible both within individual agents and between different agents.

6.3 THE AGENTS ARCHITECTURE

The architecture of the AGENTS system is simple, as shown in figure 6.1. It consists of two major components: the community of agents and the blackboard. Although AGENTS qualifies as a simple shell for developing knowledge-based systems, it may appear that it is possible to eliminate features which are generally considered essential in terms of the architecture of autonomous knowledge-based systems. For example, the knowledge base, the inference engine, a user interface, and capability for knowledge acquisition, etc., are not explicitly represented in figure 6.1. Intuitively, it might be expected that the requirements of a more sophisticated and more general problem-solving environment would require at least an equally complex architecture.

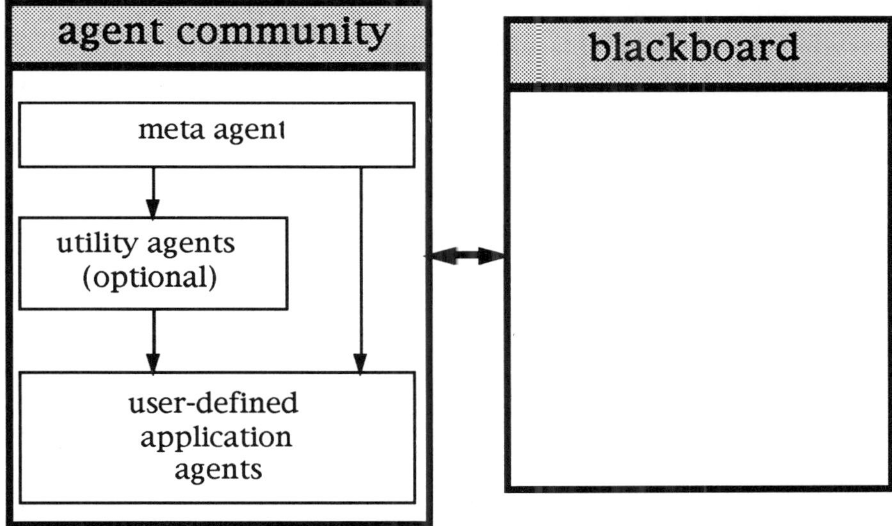

Fig. 6.1 architecture of the AGENTS shell

The reason for this apparent omission is that each individual agent in the community has its own internal architecture equivalent to that of a standard knowledge-based system; i.e. every agent has a knowledge base, an inference engine, user interface, knowledge acquisition, with an additional structure which is necessary to enable message passing for communication. The agent-based architecture is very flexible. New agents can be added into the system or obsolete agents can be dismissed from the community dynamically.

As described earlier, the community of agents comprises the set of all the agents in the system. There is one special built-in agent called the "meta" agent which is inherited by all other agents automatically. It provides a set of built-in predicates, some of which are standard Prolog predicates and some of which are extended Object-Oriented constructs. Another useful feature, albeit optional, is that there may a number of generic agents. They may be designed to provide some utility predicates, perhaps domain or problem specific, which are likely to be used frequently by more specialised agents.

Agents in the community are naturally organised into a hierarchical network. This follows from the general dependence on inheritance which implies a hierarchical structure, although the purest form of hierarchical structure – the tree – is excluded by the acceptance of multiple inheritances. Thus an agent can inherit from, or be inherited by, a number of agents, leading to a possibility of non-unique paths down the hierarchy.

The blackboard structure has the function of storing decisions in the form of nodes which constitute a network. Nodes are created, and a decision network is generated, during the process of satisfying a user goal. Such a network can be used for many purposes, for example in explaining why some particular information is required at a specific node or how a particular decision has been reached.

6.3.1 The Blackboard Metaphor

The concept underlying blackboard systems *(Englemore and Morgan 1988)* is that of a collection of experts congregating round a blackboard in order to solve a problem of common interest, as shown in figure 6.2. Participating experts (agents) interact through the medium of the blackboard by writing on, and reading from, the blackboard (cf. message passing between agents). Individual agents have the ability to determine for themselves when they can exploit items on the blackboard and how to behave. In this way, a solution to a problem is gradually built up on the blackboard by cooperating agents.

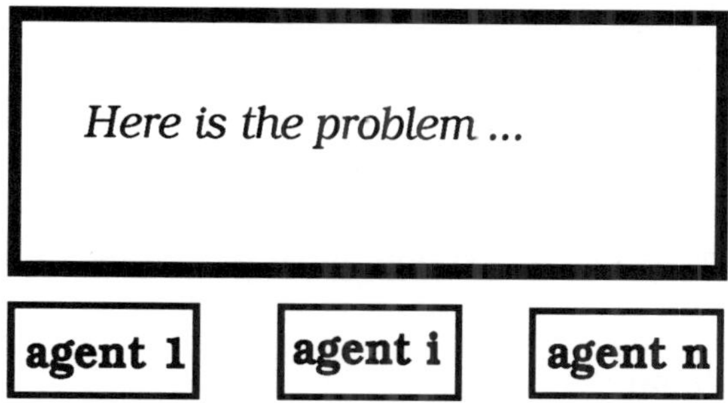

Fig. 6.2 conceptual blackboard structure

6.3.2 Blackboard Implementation

In the AGENTS system, the blackboard is implemented simply by construction of a global variable. It contains a list of data entries corresponding to the nodes which (partially) represent the problem statement. A node is a form of data model, similar to that used for nodes in the Truth Maintenance System *(de Kleer 1986 and Doyle 1980)*, for maintaining information related to a decision. The attributes which are used for describing nodes are:

- **The unique index:** A node is uniquely identifiable within the blackboard. This is achieved by assigning different numbers to nodes, starting from zero. The zero node represents the user question.

- **The datum entry:** A node usually represents a decision which is, in turn, represented in the form of Prolog terms. Variables within the datum entry must be instantiated under a specific situation.

- **The type of node:** There are four types of node including:
    - fact   – was found as a fact in an agent.
    - rule   – was derived by some other nodes.
    - user   – was specified by the user.
    - computed   – was computed by a built-in predicate.

- **The contributor of the node:** A node is proved by an agent.

- **The status of the node:** This information is not yet used in current implementation but reserved for further expansion.

- **Supported nodes:** These are the nodes whose truth value depends on the current node.
- **A list of supporting nodes:** The truth value of the current node is dependent on those of these nodes.

The two basic operations most often used in blackboard management are:

- to write an item node onto the blackboard.
- to wipe an item node off the blackboard.

To write (add) an item node onto, or to delete it from, the blackboard is simple: by adding to, or deleting from, a list. However, it is necessary to consider any implications for modifying relationships between the existing nodes by the addition of a new node or deletion of an old node, since the blackboard is a constrained decision network.

Special codes have been implanted for constructing a node for a clause, within the compiled clause procedure. Whenever a clause procedure is being run, appropriate actions are taken as far as the corresponding nodes are concerned, after variable unification and pattern matching. Typical activities involved in this are:

- To save the current environment as local to the current clause;
- To get the clause node from the supporting nodes of the current node;
- To set the current clause node as the current node; and
- To declare the supporting nodes for the current clause node.

The generation of nodes is further complicated by automatic backtracking if failure occurs. In this event, certain actions must be taken to undo those performed previously: to delete the clause node and its associated nodes from the blackboard and to restore the previously current node and node index count number. If the clause succeeds, the previously current node is reset to be the current node continuing in the success state.

6.3.3 Utilities Provided by the Blackboard

- ***Description of the Global State:*** The blackboard records the global state of the problem: what is currently accepted as true by all the agents (either by positive assertion or passive acceptance). Anything believed in the blackboard must be believed by members of the community.

- ***Recording Dependencies:*** Decisions are recorded in the blackboard in the form of nodes. The binary relationships between nodes are also retained within the node data structure. Dependencies can be constructed from the binary adjacencies whenever this is needed. They are used only for the purpose of explanation. The binary relations between nodes define the dependencies used in tracing the decision tree from the top question to the leaf facts; or a partial path from a leaf node to the root node. The former is useful for "how" explanation and the latter for "why" explanation.

- ***Truth Maintenance:*** Prolog predicates/clauses do not represent the actual beliefs but rather contain sufficient knowledge to assert a belief. Consequently, inconsistencies between predicates, or even clauses, are allowable. However, during the process of drawing an inference or proving a goal, inconsistency must be avoided. Information and knowledge contained in different agents may be inconsistent and even the information and knowledge contained in individual agents may also be inconsistent, as there is no built-in consistency maintenance mechanism in Prolog. For example, a clause GOAL is allowed to exist together with its negated form not(GOAL) in an agent world. However, such an inconsistency must be avoided in the blackboard itself if it is viewed as the decision space because a self-conflicting decision space cannot be construed as valid. The blackboard can be used for truth maintenance *(Doyle 1978 and de Kleer 1986)*. Dependency records can play an essential role in conflict resolution and consistency maintenance. In a truth-maintained blackboard, the addition or deletion of any individual decision nodes may lead to changes in the problem state. This feature has not yet been implemented in AGENTS.

● ***Communication and Control:*** In addition to the history of decisions (what has been done), the blackboard can also be used as a task agenda to represent what to do next. Tasks entering the agenda are pending messages. A message can be a query made to a known agent or a query advertised to an unknown agent, in order to identify one which is able to answer the query and negotiate with it. This is a special utility of the blackboard for communication and control, which has not yet implemented.

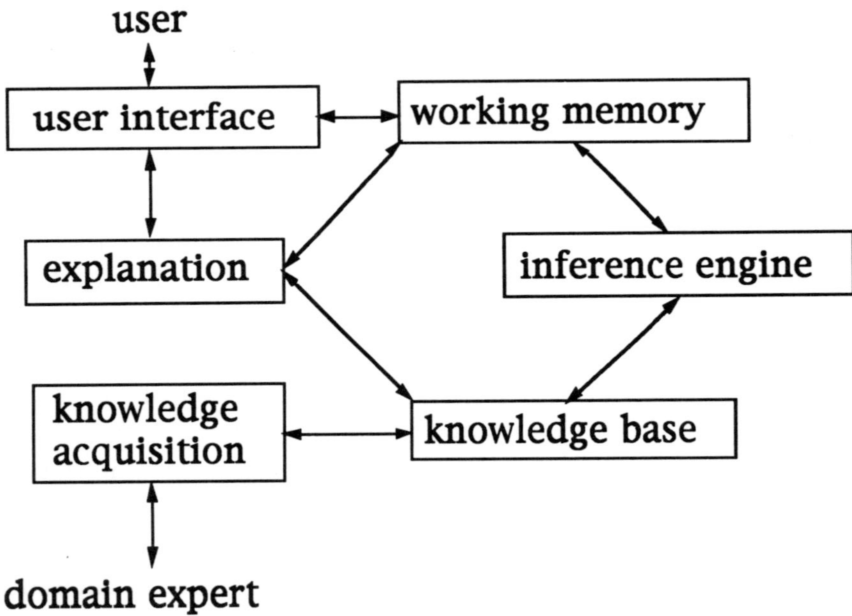

Fig. 6.3 architecture of individual agents

6.4 THE AGENT ARCHITECTURE

Knowledge-based expert systems are computer programs intended to solve problems which usually require human expertise, or to augment the decision making process of human experts. An agent has the standard structure of knowledge-based systems, as shown in figure 6.3, consisting of the following components:

(1) **Knowledge Base:** The knowledge base of an expert system is the repertoire of knowledge, containing skills, heuristics, rules of thumb, etc. Issues concerning the development of an adequate knowledge base include formal and expressive knowledge representation and dynamic learning. Facts and rules are two of the basic formalisms for knowledge representation. There are many more schemes such as semantic networks, objects, procedures, etc, which are all proven in practice.

(2) **Working Memory:** The working memory (also called the context) is a collection of descriptions of the current state of problem solving. The current state is the initial state plus those changes brought about during the execution of particular programs.

(3) **Inference Engine:** The inference engine prescribes the way in which a specific piece of knowledge is used. In general, the inference engine tends to be designed to be domain independent although domain dependent and problem specific knowledge can also be managed. This is usually achieved by devising suitable formalisms for knowledge representation and strategies for controlling inference.

(4) **User Interface and Explanation:** The majority of current expert systems are designed to be interactive: a consultation process between clients (users) and consultants (expert systems). In consequence, systems must provide facilities to facilitate user-machine interaction. A nearly-standard facility for knowledge-based systems is explanation. Two of the most common explanation facilities are "how" explanation and "why" explanation.

(5) **Knowledge Acquisition:** This facility maintains up-to-date information within the system's knowledge base so that up-to-date decisions can be traced.

Most knowledge-based expert systems are designed to be interactive. That is, (at least) two agents are primarily involved: the user and the system. They demand help from each other. It has been argued *(Forsyth 1986)* that the user interface is not an optional component in a knowledge-based expert system. The rest of this section is concerned with interaction between the user and the system within the Object-Oriented Prolog (AGENTS) environment.

6.4.1 User Modelling

User modelling is of great significance, not only in systems that have the restricted role of supporting problem solving, but also systems which are intended to be sufficiently powerful to solve problems primarily without external intervention, since problem solving, in general, cannot be fully automated, especially when the problem is complicated. Two of the main roles of the user are:

- To provide specific information needed by the system.
- To control the process when the system comes to a crossroad.

The role of the user within an agent-based environment is complicated by the fact that the user is only one, albeit special, agent among the agent community. If the user is supposed to play the above two roles, (s)he must be multi-disciplined because of interactions with several specialist agents.

In addition to the "meta" agent, there can be another built-in agent called the "user" agent. As the "meta" agent is inherited by all the other system agents, the "user" agent is also inherited by all the system agents by default. The use of the "user" agent is significant in computer assisted problem solving involving multiple agents, because the human user can always be considered as one member within the agent community. As in most knowledge-based expert systems, commands from the user have the highest priority for execution. On the other hand, user interventions have the lowest priority. For example, in the case where the system has difficulty in deriving the answer to a specific question, or a conflict is encountered, the user would be consulted only as a last resort.

6.4.2 User Interview

Most existing expert systems initiate their activities with an interview with the user to acquire domain-dependent and problem specific information. The process of interview also serves as a process of learning. For some types of problem solving such as diagnosis, a solution to a problem is generated after an interview is finished. The process of problem solving proceeds according to the structures which are programmed in the system and is assisted by user answers. Most of the early knowledge-based systems for mechanical design were also constructed in such a fashion.

Problem specifications are usually early-committed. That is, anything necessary for problem solution is asked by the system. On the other hand, decision making is, in general, least-committed. That is, information for making a specific decision is requested whenever it is required by the system. Opportunistic commitment applies to the user interview. Whenever the system believes something helpful, it addresses appropriate enquiries to the user. The early-committed user interview may be designed as a procedure to ask a sequence of questions. The least-committed user interview may be designed as a procedure for the user to intervene in the system's decision making activities. This sub-section is limited to the discussion on opportunistically-committed user interviews: asking for available information to satisfy current system needs. The system searches for a predicate which matches the current goal. Then, clauses within the predicate are tested one by one in the order that they were defined. The clause under consideration is tested to see whether it is what the system needs in order to prove the goal.

There is a built-in predicate ?- for defining askable goals. It takes one argument (cf. ?- has two arguments when it is used as the message-passing mechanism). The truth value and associated variables of the askable clause must be provided by the user. Askable clauses are compiled into procedures which are also retained within the Prolog procedure pile corresponding to the predicate. Note the significance of the order in which askable clauses are placed. The effect is exactly the same as that of ordinary Prolog clauses. The backtracking mechanism will test each one in the order they are defined.

This procedure deals with interactions between the system and the user. It prints the goal for the user and then waits for an appropriate user input. There are three options for the user to react concerning the goal being asked:

- "why"   – User can ask system why this goal is needed.
- "yes"   – The user confirms that the goal is true.
- "no"   – The user has confirmed that the goal fails.

The user may respond with the "why" query, which will then invoke the system's "why" explanation facility to show the reason that this goal clause is useful. The user may confirm that the goal clause is true by responding "yes"; or refute that the goal clause is false by responding "no". If the user answer is "no", the askable goal fails and backtracking takes place.

If the user answered "yes", further questions may be raised relating to uninstantiated variables contained in the goal clause. The system will prompt the user for a value for each variable. The user may choose either to provide a value for the variable or to leave it undefined. The exploration then proceeds. Occasionally, it may be necessary for the user to refute the goal under consideration on being asked for variable values at this stage.

6.4.3 "Why" Explanation

An expert system behaves like a human expert for solving a particular problem. It also should be capable of explaining its decisions ("how" explanation) and the underlying reasoning or intentions at appropriate stages ("why" explanation). Both "how" and "why" explanation facilities are based on dependencies recorded in the blackboard.

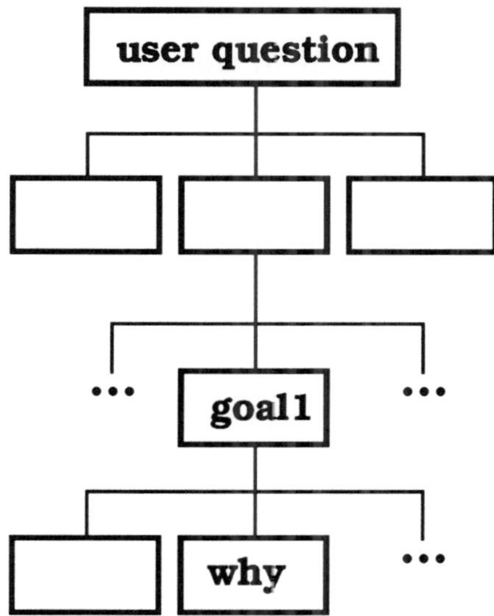

Fig. 6.4 "why" explanation

The "why" explanation can only be invoked during an interactive session involving the user at the time when a goal is askable: i.e. the system asks the user for some information but the user wants to know why this information is particularly needed: for example,

- why is this piece of information useful?
- why do you do that?

The system will respond with explanation looking like this:

    GOAL: [goal] CAN BE USED TO PROVE:
    GOAL: [goal] CAN BE USED TO PROVE:
    GOAL: [goal] WAS YOUR ORIGINAL QUESTION.

In practice, the "why" explanation facility operates by showing a tracing path from the current goal (which is being asked for the user) to the top goal (which is the original question from the user), as shown in figure 6.4.

6.4.4 "How" Explanation

Once the system has produced an answer to the user's question, the user may enquire how this conclusion was reached:

- How did you arrive at this conclusion?
- How did you do that?

The system may (typically) respond:

YOUR QUESTION: [question] WAS PROVED BY:
GOAL: [goal] WAS DERIVED BY:
GOAL: [goal] WAS FOUND AS FACT; AND
GOAL: [goal] WAS TOLD BY USER.

The "how" explanation facility provides the user with a tree-like trace from the top goal node to its root node. The trace for a "how" explanation is the same as that for a "why" explanation, provided by the blackboard. The difference between "why" explanation and "how" explanation lies in that the "why" explanation shows the user a partial path from the current goal to the user goal while the "how" explanation show the user the complete trace tree, as shown in figure 6.5.

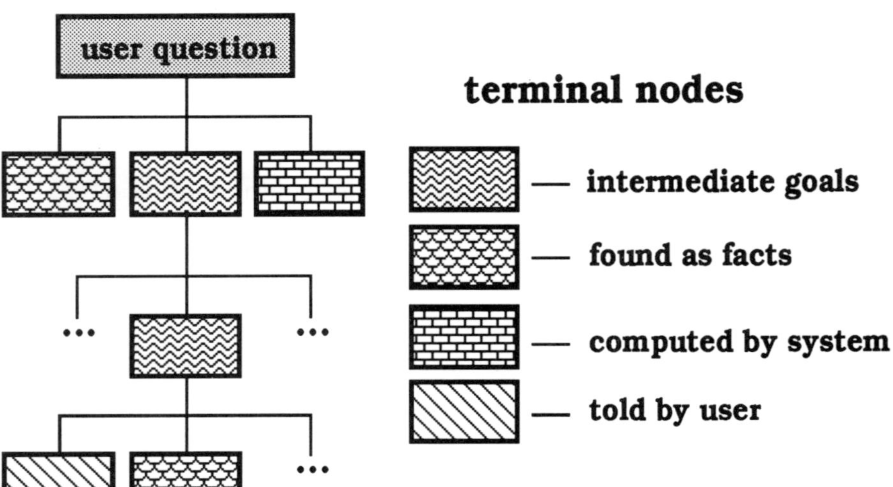

Fig. 6.5 "how" explanation

The "how" explanation is a part of the top-level interpreter of the AGENTS system. No "how" explanation is provided if a user question is not proved. If the user question is proved, the system providesthree options for the user:

- "how" – User may ask system to explain how solution is reached.
- "more" – User may ask system to generate more solutions.
- "exit" – User is satisfied with solution and exit system.

Another potential application of the "how" explanation is for proposal assessment, i.e. to assess the merit of a solution. This is more complicated than explanation of how a solution is reached since proposal assessment depends on both the constitution of a solution and its process of generation (see Chapters 3 and 4 for more detailed theoretical discussions). There is some justification for suggesting that proposal assessment should not be grouped under the category of explanation but rather as a type of strategy for decision making. It should be noted that "how" explanation is only concerned with explaining successful decisions and not for failed attempts. Reasoning about failures is, however, useful:

- To make a suggestion for further alternative attempts.
- To avoid any futile attempts that have been proved to be unacceptable.
- To compromise a failure to obtain a nearly satisfactory solution.

6.5 AGENTS DEVELOPMENT

6.5.1 Prolog Development

Prolog provides a small, but useful, set of mechanisms for developing knowledge-based systems. They include:

- Rule-based knowledge representations
- Rule-based backward chained inference
- Pattern unification to support clause-based declarative programming
- The ability to have suspended variable bindings via the logical

variables

- Simple and elegant semantics that offer a cohesive framework for rule-based systems and allow the formal manipulation of logical programs

Most of the above features can be inherited directly by a knowledge–based shell. Indeed, Prolog is already almost a knowledge-based programming shell. However, there are important desirable features such as explanation and forward chained inference which are not supported in Prolog. To incorporate these features, there are some special requirements, specifically:

- need for a top-level shell interpreter.
- need for special formalisms for representing knowledge.

A Prolog program can be considered to consist of rules (including facts which are rules that always holds unconditionally). Unfortunately, this kind of built-in rule-based feature must be adapted to support a knowledge-based system, so as to include an explanation facility and forward chained inference. It is worthwhile to point out that most of the current expert systems implemented in Prolog have special formalisms for representing domain knowledge and, therefore, special inference engines are developed to suit these formalisms. Take, for example, the architectural expert design system BUILD *(Roseman et al 1985)*, where rules are represented in the following form:

ruleNumber(Head, Body).

As another example, *Bratko (1987)* demonstrated an expert system shell in Prolog, where facts are represented of the form:

fact : FactBody.

whilst rules are represented as:

RuleIndex : if Condition then Goal.

Special formalisms for rule representation are able to support inference involving both backward and forward chaining, and to provide an explanation facility. However, the built-in Prolog interpreter (inference engine) is no longer able to make use of rules in the required fashion. Special procedures are then required to interpret them. For example, the interpretation procedure - **explore-** *(Bratko 1987)* is extremely similar to the built-in Prolog inference engine, with some extensions such as user interaction. Hence it is able to perform forward chained inference. As Bratko pointed out, such an interpretation written in Prolog is indeed more procedurally biased than usual, while the declarative feature seems to be sacrificed. However, the declarative clarity still remains as a major advantage in terms of knowledge representation of facts and rules.

It is interesting to notice that system developers do not make direct use of Prolog facts and rules as the representation formalism, and the built-in Prolog inference engine as the major inference mechanism, if they are determined to select Prolog for their implementation of domain expert systems. The reason may be that they are limited by facilities provided by Prolog, though, as a generic tool, there is still great flexibility for constructing systems or integrating with other facilities such as blackboards, for specific purposes. With the AGENTS system, there is no need for any special formalisms of representation or any special inference engines, while most of the desirable knowledge – based facilities such as explanations can be achieved and have been supplied within the system.

6.5.2 Knowledge Representations

The declarative syntax of Prolog programs, in conjunction with procedural operational semantics, allows the combination of declarative and procedural representations of knowledge, within a coherent logical framework. AGENTS inherits all the Prolog features with added Object-Oriented constructs:

- facts
- rules
- attributes
- inheritance through abstraction pointers

- constraints
- objects

In the above set of constructs, typical built-in structural facilities for knowledge representation are listed. System developers are as free as in Prolog to design special or complicated formalisms derived from the basic constructs. In Chapter 7, two special purpose representations of the constraint network and the Pros-and-Cons model will be discussed. Other advanced representation schemes in Prolog have been discussed in a number of texts *(see for example Bratko 1986)*.

AGENTS rules are exactly the same as Prolog rules:

Head :- Body.

where

(1) Head – provides a pattern based on which a rule is considered.
(2) Body –   represents a set of sub-goals to be proved one after another.

In AGENTS, sub-goals in the rule body have the role of messages to be sent to appropriate receiving agents. There are two logical ways of interpreting rules of the above form. The first is to consider the Head as a condition and the Body as an action:

if Head then Body.

Another is to consider the Head as the conclusion and Body as the corresponding condition:

if Body then Head.

Unlike built-in Prolog rules, AGENTS rules can be used directly to provide an explanation facility without the need for an intermediate inference engine.

Both attributes and facts can be used to represent the internal state of an agent. They are, after all, different types of agent properties:

(1) Attributes are originally a kind of Object-Oriented construct,

while facts are a special kind of rules without conditions.

(2) In general, it is advantageous to use attributes to represent relatively static features or characteristics possessed by an agent. On the other hand, facts would be used for comparatively dynamic agent properties.

(3) Both attributes and facts of an agent are inheritable during the process of classification. That is, sub-classes inherit attributes and facts from their super-classes.

(4) Both attributes and facts are inheritable as a consequence of instantiation. Objects inherit attributes and default values from the agent explicitly. However, objects inherit methods (facts are special methods) implicitly.

Cooperation between agents is achieved through sharing of information. In AGENTS, there are two types of information sharing: inheritance and message passing. Their relative significance and interaction can be appreciated by consideration of the following points:

(1) Both inheritance and message passing deal with inter-agent sharing of properties. However, inheritance specifies the sharing of whole properties of all kinds (both attributes and methods). Message passing specifies the sharing of one method / attribute at a time.

(2) Inheritance implies static sharing of information, whereas message passing enables dynamic sharing. Inheritance pointers specify structural inter-agent relationships which are relatively static (though changeable at run-time). Message passing always takes place at the run-time.

Chapter 3 discussed a number of abstraction operations commonly used in product design and modelling for complexity management. The AGENTS system does not, however, provide constructs ready for all of them. Nevertheless, it is possible to achieve abstraction such as configuration and projection using basic AGENTS constructs.

The AGENTS system provides a useful facility of constraints: active clauses to be executed when certain actions are taken to some agent properties. It should be noted that conditions of a rule are also called (rule) constraints. Consider a primary clause with constraints of the following form:

```
Head =: Before.          – before constraint
Head :- Body.            – primary method
Head := After.           – after constraint
```

This is functionally equivalent to a single clause:

```
Head :- Before, Body, After.
```

The treatment of constraints in the AGENTS system is influenced by the following factors:

(1) The advantage of breaking a constrained clause into three components of Before, Body and After is that constraints can be declaratively added to or deleted from the program without affecting the primary clause. The penalty is that there are three stages of head pattern matching.

(2) Demon constraints are imposed on predicates (and are also active for facts) while rule constraints are associated with specific rules (factual clauses have null constraints).

(3) Demon constraints can be defined dynamically, and attached to predicates without affecting the primary predicate clauses. A predicate can have as many as constraints as necessary. On the other hand, addition of any conditions to a rule is a process of re-defining a rule - deleting the old one and constructing a new one.

6.5.3 Process of Inference

The model that Prolog makes inferences is characterised by the following features (see also Chapter 5):

- goal-directed.
- rule-based backward chaining.
- pattern unification (matching).
- automatic backtracking.

The way that AGENTS answers questions is similar to that in which Prolog answers questions, i.e. answering a question means attempting to satisfy a list of goals. They can be satisfied if the variables that occur in the goals can be instantiated in such a way that the goals logically follow from the program. That is, a list of goals is executed with respect to a given program, as shown in figure 6.6. Note that:

(1) If the goal is satisfiable then the success indication "yes" is given; otherwise this indicates a failure.
(2) An instantiation of variables is only produced in the case of successful invocation; in the case of failure there is no variable instantiation.

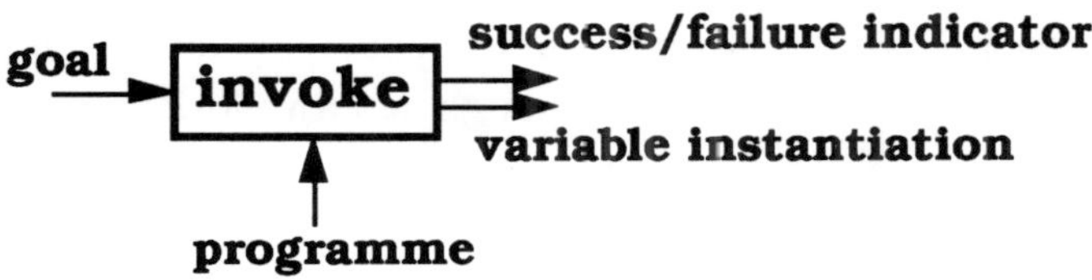

Fig. 6.6 invoke procedure

There are four possibilities that a goal Q is ultimately satisfied:

(1) Q is established as a belief in the blackboard (But not yet used in AGENTS).
(2) Q is found as a fact in the knowledge base of the agent.
(3) Q is computed by a built-in predicate.
(4) Q is an 'askable' question and answered by the user.

AGENTS provides multiple inheritance among agents, thereby avoiding redundant copying of information among agents. Instead, properties are automatically shared. For example, if there are two agents called "vehicle" and "car", defined as follows:

```
agent(vehicle).
    sub(car).
    registration(No) :-
        write(No).
```

```
endagent.
agent(car).
       super(vehicle).
endagent.
```

The "car" agent inherits from the "vehicle" agent; or the "vehicle" agent is inherited by the "car" agent. There is a clause for writing the registration number on the screen defined within the "vehicle" agent. Note that this predicate is not explicitly defined within the "car" agent. However, a message to the "car" agent in the following request will also print the registration number of the car on the screen through the inheritance:

```
car ?- registration(Number).
```

In general, an agent answers a question by asking appropriate questions to relevant agents, although it can work independently. Consider an example of a kind of cooperation:

```
cooperation([], _).
cooperation(_, []).
cooperation([H1 | T1], [H2 | T2]) :-
       H1 ?- H2,
       cooperation(T1, T2).
```

where

(1) The first argument is a list of agents and the second is a list of tasks.
(2) The cooperation succeeds if either or both of the argument lists become empty.
(3) Each agent is sent a corresponding task message one by one until success.
(4) The cooperation fails if any of agents fails to complete a task.

Consider another example of cooperation:

```
cooperation([], _) :- fail.
cooperation([H | T], Task) :-
```

```
        H ?- Task.
   cooperation([H|T], Task) :-
        cooperation(T, Task).
```

where

(1) The first argument is a list of agents and the second is a task.

(2) The cooperation fails if none of the agents can complete the task.

(3) The cooperation succeeds if one of the agents can complete the task.

(4) Another agent is tried after one fails till either success or failure occurs.

The introduction of demons is one the most important features of Object-Oriented Prolog, since they have a large effect on the process of reasoning, as was discussed in the last chapter. Here, it is interesting to enquire what will happen if a constrained clause is defined recursively. Consider a very simple program which is defined within the agent named "funny" as follows:

```
agent(funny).
    explore :-
            write("What's so funny?"), nl,
            explore.
    explore.
    explore := constraint.
    constraint :- write("Interesting!").
endagent.
```

When the "explore" message is sent to the "funny" agent:

```
funny ?- explore.
```

The "funny" agent executes its "explore" clause method which will return true. Due to the defined constraint on the "explore" procedure, the output will look like:

```
Interesting! ?
```

If backtracking is forced, the output looks like

> What's so funny?
> Interesting!Interesting!
> It's funny!

Two Interesting!'s appear together. The first is due to the execution of "explore" within the recursive rule and the second is caused by the recursive rule itself. Is this type of feature what a programmer needs? Possibly not - but it exists anyway as an indicator that a recursive constraint has been applied. Particular attention must be paid when a decision is made to use a constraint on a recursive predicate.

6.6 SUMMARY

This chapter has discussed the general principles of the programming techniques used in AGENTS, with particular reference to the corresponding constructs in standard Prolog. These features can be summarised as follows:

(1) The AGENTS shell does not require any special formalisms for representing knowledge as it utilises ordinary Prolog clauses.
(2) The shell does not need a special inference engine beyond that which is already built into standard Prolog.
(3) The shell is able to explain intentions and decisions without any special requirements relating to knowledge representation or inference making, based on dependencies recorded in the blackboard.

The above features provide an environment which is familiar to developers who are already familiar with programming in Prolog. In addition it is of considerable value for designers who are constrained in their choice of programming environment by compatibility specifications, especially where these are imposed by third party commitments. With this approach it becomes possible to design significantly more complex expert systems, particularly for engineering design, since it is unnecessary to design special knowledge - representation formalisms or inference engines or explanations.

Chapter 7
Design with AGENTS

7.1 INTRODUCTION

The AGENTS system provides a tool for constructing cooperating knowledge-based systems. It possesses a set of basic constructs for distributed knowledge representation and an inference engine for drawing collective inferences. Compared with the theoretical design framework developed in Chapters 3 and 4, the AGENTS system still requires enhancements to incorporate elements which are essential, and available elsewhere, to qualify as a practical design platform. Nevertheless, the current prototype system is sufficient for demonstrating its utility in manufacturing design, with more ready-for-use and easy-to-use facilities than ordinary Prolog.

This chapter concentrates on cooperative design problem solving using AGENTS, although its applicability is considerably more general. It focuses on two types of design problem:

- morphological conceptual design
- constrained layout configuration.

The primary objective is to develop appropriate design engines which make use of formalisms suitable for eliciting knowledge and compatible algorithms suitable for making design decisions.

The next section discusses three typical design methods for three classes of design problem solving. In section 3, the well-known "robot-blocks" problem is used to demonstrate basic ideas of agent-based design problem solving. Section 4 is concerned with agent-based morphological conceptual design determining the composition of an artifact. Section 5 presents a technique of constraint representation for configuring a layout from the corresponding composition.

7.2 DESIGN METHODS

7.2.1 Classes of Design Problems

Design problems fall into certain classes according to their complexity. In the present work the following classification will be used:

Class 1: A design problem is of class 1 if the generic structure of an artifact is unknown.

Class 2: A design problem is of class 2 if the generic structure of an artifact is known but the specific scheme (composition and layout) is unknown.

Class 3: A design problem is of class 3 if both the composition and layout of an artifact are known.

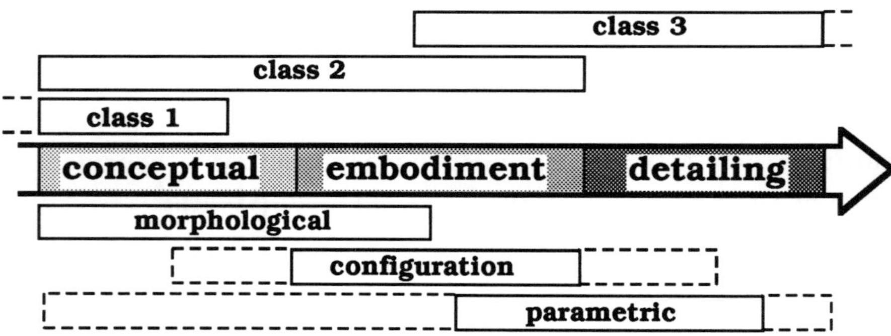

Fig. 7.1 classes of problem, classes of method, and processes of design

This classification is similar to that proposed by *Brown (1984).* A principal distinction is that the classification here uses the concept of generic and schematic networks. Traditional CAD systems deal with design problems within class 3, while expert systems can extend CAD up to the vague boundary between class 3 problems and class 2 design. In domains such as manufacturing design, some problems in class 2 can be dealt with by cooperating expert systems. Indeed, this is the main theme of the current volume, i.e. from the generic structure of a design problem it is possible to determine a suitable scheme, including its composition and layout, although substantial difficulties may exist in achieving this goal. In addition, it is important to note that there are no clear and sharp boundaries between these three classes of design problem.

Throughout the design process and across the spectrum of design problems there are three general classes of design methods widely used:

- parametric (variational) design
- configuration design
- morphological design.

In general, routine design of class 3 is parametric at the detailing stage and possibly the latter stages of embodiment. Configuration is the primary activity of embodiment design where the main aim is to establish the scheme for a set of components. It corresponds approximately to a design problem of class 2. The major task of conceptual design is to obtain a set of components which constitute an artifact. A morphological method is often used for this class 2 design problem. Design problems of class 1 are not included for discussion because their complexity and uncertainty are beyond the scope for computer systems to manage. Figure 7.1 depicts the approximate correspondences between design processes, design classes, and design methods.

7.2.2 Morphological Conceptual Design

Over the years morphological approaches have played an important role as a general aid to design, especially in the early stages such as conceptual design. These approaches are based on developing a list of functional requirements and a table of options for implementing functions. An embryo design is any realisable combination of functions and structures. However, an acceptable design must satisfy some constraints, both domain dependent and independent.

The method of morphological conceptual design starts from a list of key functional requirements and a primitive table of options, and proceeds under a set of constraints to a state where all functions are implemented and no constraints are violated. Following the principles of General Design Theory *(Yoshikawa 1981)*, conceptual design is concerned with the following three types of task:

(1) **Functional Reasoning:** This is mainly concerned with the processing of functional requirements from raw input information. This activity is usually both problem specific and domain dependent. In addition, it is primarily heuristic, and therefore is usually accomplished by human designers.

(2) **Structural Reasoning:** This deals with the establishment of the table of optional structures for fulfilling functions. It is, in general, problem independent but domain specific. That is to say, modular entities with both functional and structural descriptions can be identified without considering a specific problem in detail.

(3) **Mapping (Transformation):** Mapping between the requirement list and option table involves the selection of one option in the table and attempts to implement a functional requirement. Such a mapping requires overall consideration of global constraints while satisfying imposed or local constraints.

It has been observed that the mapping between functional and structural spaces is nearly isomorphic in most circumstances. That is, a function is uniquely implemented by a structural feature. However, local distortion may well exist. In other words, a single structural entity may implement more than one function, or a single function may be implemented by more than one structure, or structures are shared by different functions. This is called non-isomorphic mapping. Two types of non-isomorphic mapping are useful *(Tong 1987)*:

(1) ***structure-sharing:*** Structure-sharing exploits the fact that a single entity (structural or functional) is capable of serving more than one purpose, either at a single time or at different times.
(2) ***structure-merging:*** Structure-merging detects functional redundancies and removes them by merging their implementations.

It should be noted that structure-merging occurs between two components being used for the same purpose at higher levels while structure-sharing occurs when the same component is used for two different purposes.

7.2.3 Configuration Design

Configuration design is concerned with the establishment of relationships, especially spatial ones, between the components of a product, to obtain a general scheme under which they are arranged. It is not specifically concerned with assignment of specific values to attributes describing individual components, although this assignment may take place, consequentially, during the process of configuration.

Configuration usually succeeds conceptual design and commonly predominates during embodiment design. However, an early decision made at conceptual design for the selection of an appropriate structure corresponding to a given function must also take into account its position within the product. Likewise, a decision at the detailing stage of a component can also have effects on the relative position between itself and other components; i.e. consideration of configuration at the conceptual and detailing levels is often necessary.

The configuration problem can be formulated as a process of generation of a (constraint) network which represents the state of a problem solution. The configuration starts with a set of initial objects (components),some of whose relationships may have already be en   partially described. It then proceeds by establishing further relationships between objects. The whole process of configuration is subject to constraints which may be local to components or global to the network (problem). An evaluation function, for instance a cost objective, can also be assessed.

Configuration design during the embodiment stage can in turn be divided into two cases. The first concerns pure configuration. This does not deal with abstractions such as formation (of new concepts), or mutation (of local components); i.e. pure configuration assumes that the composition of a design artifact is already known and does not change during the process of configuration. In the second case, some of the major functional requirements remain to be clarified and components are generalised and are liable to change with the progress of configuration. Their spatial relationships, some leading dimensions and required duties are to be settled.

7.2.4 Parametric Design

In parametric design, the generic and schematic structures of the design artifact are commonly predetermined and are described in terms of attributes or features. It involves assigning values to attributes, called parametric design variables *(Deitz 1989, and Drake and Fela 1989)*. The values to be assigned are not necessarily numeric. They may also be symbolic. Parametric design usually takes an analytical approach because design variables can be described by equations or inequalities. The entire design optimisation problem can be usually determined by evaluating a single objective function (minimisation or maximisation). The idea of parametric design has also been introduced in geometric design *(Deitz 1989)*. It is alternatively described as feature-based design or variational design *(Suzuki et al 1986)*.

One of the crucial, but difficult, issues of truly parametric systems is the construction of a suitable data structure, so that users and systems can modify a design as easily and quickly as it was created initially. That is, the data model must be robust enough to cope with the evolutionary nature of design from creation, through maintenance, modification and final acceptance. One advantage that parametric design offers is the capability to re-use previous designs by varying values of their parameters.

Parametric design can also be used at the conceptual level if entities in both the functional space and the structural space can be described in terms of pre-identified attributes. Configuration can be achieved by a parametric process if the structure of the network is determined and only parameters may vary. Parametric design methods should be utilised wherever possible since techniques and algorithms are well established.

7.3 AGENT-BASED DESIGN: THE ROBOT-BLOCKS EXAMPLE

Some simple but famous problems such as the "Robot-Blocks" problem and the "Monkey-Bananas" problem have long been used to demonstrate problem–solving strategies in Artificial Intelligence. The "Robot-Blocks" problem is used here to illustrate the basic ideas of layout configuration of a machine tool at a coarse granularity, by visualising the ROBOT as a designer and the blocks as machine components.

7.3.1 The "Robot-Blocks" Problem

The "robot-blocks" problem can be described as follows: The WORLD consists of a ROBOT, a DESK, and a number of BLOCKS which are placed on SHELF. To satisfy the rules of the GAME, the ROBOT is required to construct an assembly by using its HAND to take each BLOCK off the SHELF and put it onto the DESK or assemble it with other BLOCK(S) already on the DESK.

For example, the ROBOT is required to construct a game machining centre as shown in figure 7.2. To achieve the goal, the ROBOT does the following two things:

(1) To select relevant BLOCKs included in the GAME 'machine' from the SHELF and place them on the DESK, the process known as the process of embodiment.

(2) To arrange the BLOCKs on the DESK into the required structure

which conforms to the objectives of the GAME, the activity described as the process of configuration.

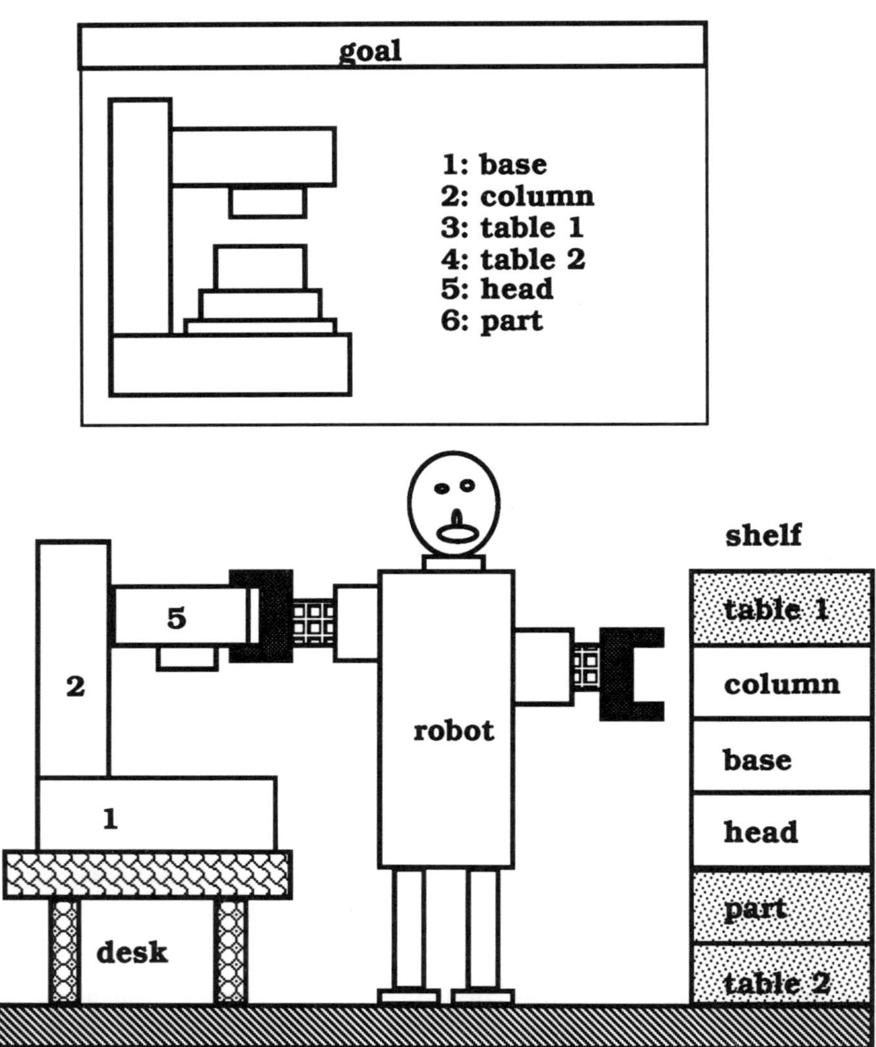

Fig. 7.2 game: ROBOT builds MACHINE

The BLOCKS are simply referred by their names: Spindle Head (HEAD), Base (BASE), Column (COLUMN), Table 1 (TABLE 1), Table 2 (TABLE 2), DESK, and ROBOT's hand (HAND). The arrangement of BLOCKS when they are assembled on the DESK is significant. For example, the initial state of the problem may be described as follows (in English descriptions):

DESK is empty and all the BLOCKS are on SHELF.

Informal English-like descriptions of the final state (goal) in figure 7.2 of the problem may include the following:

- BASE is on the DESK.
- COLUMN is on the top of BASE to the left-hand side.
- TABLE 1 is on the top of BASE to the right-hand side.
- TABLE 2 is on the top of TABLE 1.
- PART is on the TABLE 2 and under HEAD.
- HEAD is attached to the right-hand side of COLUMN.

Intermediate states of the world change dynamically, leading to either a successful plan or a failure. For example, the partial state of the world in figure 7.2 can be described by the following English sentences:

- BASE is on the DESK.
- COLUMN is on the top of BASE to the left-hand side.
- HAND is holding the HEAD.
- TABLE 1 is on the SHELF.
- TABLE 2 is on the SHELF.
- PART is on the SHELF.

The above descriptions are fairly coarse-grained:

(1) It does not give descriptions of individual objects.
(2) It does not involve any specifications of absolute positions.
(3) The relationships between objects are also roughly described.

Note that there may be a large number of statements about the world which are not mentioned in this description. For those such as:

"AGENTS is a prototype Object-Oriented Prolog",

this seems irrelevant to the ROBOT's operations. On the other hand, for those such as:

"COLUMN has already been put on BASE",

or

"HEAD is on DESK",

performing a specific operation may well change their truth value.

The ROBOT is capable of doing something like "clean the top of TABLE", "pick up the COLUMN", "put down the spindle HEAD", etc. These operations are not those often used in design but may be used in assembly. Thinking about the operation of picking up the block HEAD, it can now be described in English as follows:

The block HEAD may be picked up if it is clear and the ROBOT's hand is empty. When it has been picked up, then the ROBOT's hand will be holding it. The block HEAD will no longer be on the surface it was on before the operation was performed.

It can be seen that the activity of the ROBOT depends on:

- PRECONDITIONS that must be true for OPERATION to perform;
- PRIMARY OPERATIONS that accomplish   the job;
- CHANGES that the operation makes:
  - NEWLY TRUES that were not true before it was performed but are afterwards;
  - NEWLY FALSES that were true before but are not after.

The ROBOT builds the GAME 'machining centre' by following the instructions below:

- put the COLUMN on the BASE
- put the TABLE 1 on the BASE
- put the TABLE 2 on the TABLE 1
- put the HEAD on side of the COLUMN

However, what is provided to the ROBOT is not a sequence of commands as given above. Instead, the ROBOT must understand the current state of the problem and generate a plan for approaching the goal state. That is, the ROBOT actually must first understand the building GOAL given by COMMAND and then work out the PLAN by which it can ACHIEVE the required GOAL.

BLOCKS in the above "robot-blocks" problem are deliberately shaped to look like the structural modules of machining centres. However, this does not mean designing or assembling a machining centre is that simple. The insights which can be gained from this simple example are summarised as follows:

(1) The problem allows the exploration of the general approach to computer assisted problem solving. The approach used above is called means-end analysis, adopted in systems such as GPS and STRIPS *(Nilsson 1977)*. Some design systems are also based on this strategy although extensions may be made. For instance, the MOLGEN system is applied in genetic engineering, where the effects of different sequences of laboratory operations are investigated in order to produce a sequence (or sequences) which leads to the desired substance *(Stefik 1981a, b)*.

(2) In the approach to the robot-blocks example described above it is for the ROBOT to decide where a specific BLOCK should be placed, i.e. it is the ROBOT which exhibits intelligence. Another approach is to distribute intelligence to individual blocks. That is, the ROBOT inquires of each BLOCK where it should be placed. In this case, the ROBOT simply sends a message to a BLOCK:

    ROBOT: Where are you usually placed?
    COLUMN: On the top of foundation.
    ROBOT: Place column on the top of foundation.

(3) The ROBOT may consult a group of advisors for the validation of a specific action. For example the ROBOT will validate its intention of placing the column on top of the foundation:

    ROBOT: Is it legal to put a column on the top of foundation?
    ADVISOR: Yes.

That is, every decision step made by ROBOT must be justified.

Again, supposing that the ROBOT is monitored by a group of specialist advisors, each advisor may be in favour of, or against, a specific decision step. In the above example of a dialogue session between ROBOT, BLOCKS, and ADVISORS, the decision of putting COLUMN on the top of FOUNDATION is a legal action according to the advisory committee ADVISORS. On the other hand, a decision to put FOUNDATION on the top of COLUMN is refuted by ADVISORS. In fact, such a decision should never have been suggested either by FOUNDATION or COLUMN.

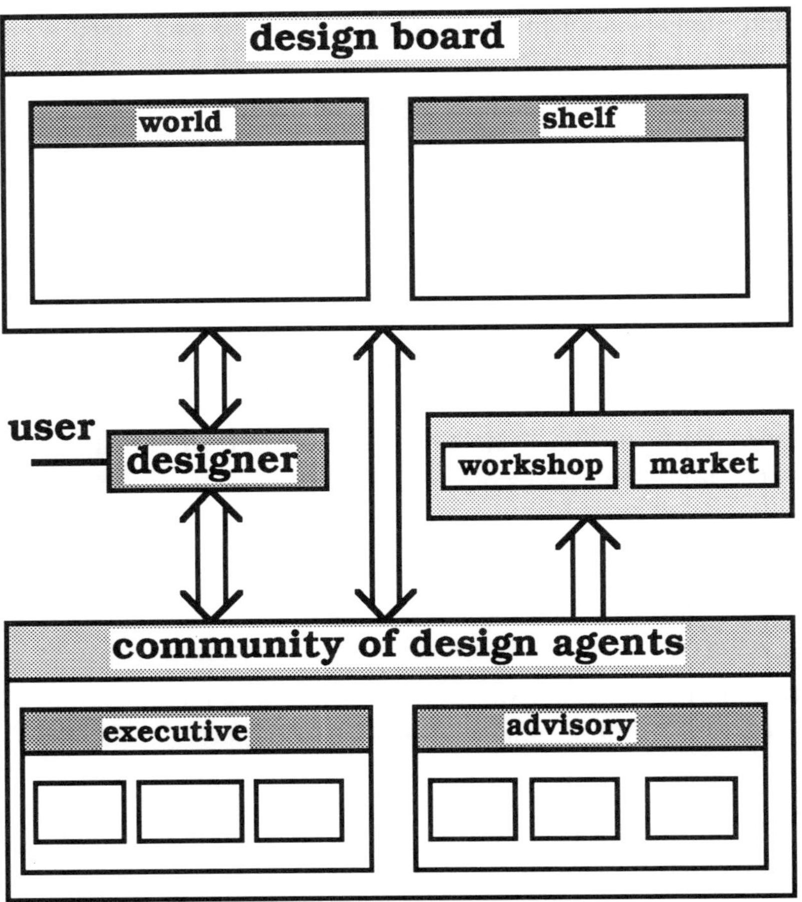

Fig. 7.3 *conceptual model of agent-based design*

7.3.2 Agent-Based Design

Figure 7.3 shows a conceptual model of agent-based cooperative design problem solving. Two of the basic elements in the architecture are the design board and the community of design agents. These two elements are coordinated by a central designer. The key role of the central designer, which may be a member in the community of agents, is to read a list of descriptions of design requirements from the user and send a message to the executive agent community to ask them to start a design session. The actual design activities are carried out by executive design agents.

There are two sections in the design board. One is the SHELF and the other is the WORLD. In general, the SHELF is used to store components which comprise the product to be designed, while the WORLD represents the scheme of the product (not only components but also its layout). The separation between the SHELF and WORLD is made because it is assumed that product design can be divided into two stages. The first stage of conceptual design is to determine the composition by morphological analysis. Then the layout is determined by constrained propagation through a network representation.

The community of design agents is also segregated into two groups, the first called the executive group of agents and the other the advisory group. An executive agent is responsible for designing a component within the product. An advisory agent is responsible for monitoring the design decisions made by an executive agent to ensure that they are satisfactory.

7.4 AGENT-BASED MORPHOLOGICAL CONCEPTUAL DESIGN

7.4.1 Basic Morphological Mapping Algorithm

Figure 7.4 shows a conceptual architecture of a morphological conceptual design system. It uses three components:

(1) **A set of requirements R:** The requirement set R can be represented by a predicate of the following form:

functional_requirements([H | T]).

(2) **A table of options E:** The option table E can be represented by the following predicate:

    option_table([H|T]).

(3) **A set of components C:** The component set C can also be represented by a predicate of the following form:

    structural_components([H|T]).

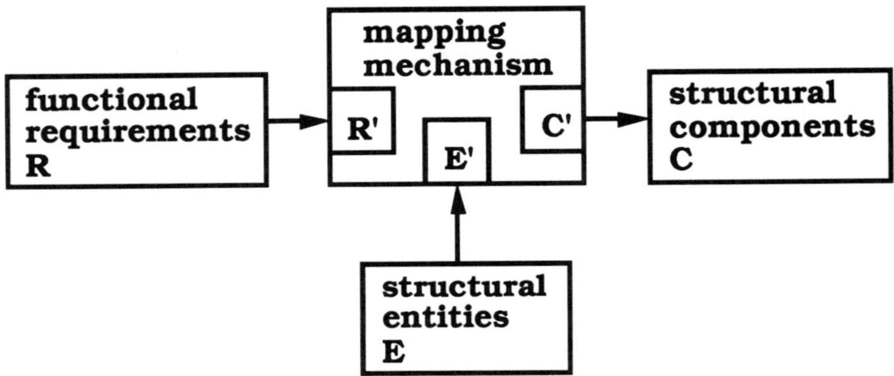

Fig. 7.4 *morphological mapping*

A component in C can be represented by either an agent or an object (an instance of an agent). The remaining discussion in this chapter assumes that components are represented by objects while members in E set are agents.

The mapping mechanism takes R as its input, E as its resources, and C as its output. It uses three local variables R', E', and C' to represent intermediate states of R, E, and C respectively. The major function can be outlined by the following algorithm:

    (1) retrieve R.
    (2) assign R to R'.
    (3) if R is empty, store C' in C and exit with success.
    (4) consider one element f in R and put the rest in R'.
    (5) retrieve E.
    (6) assign E to E'.
    (7) if E is empty, exit with failure.

(8) generate one element *s* from E and put the rest in E'.
(9) test if *s* can fulfil *f*.
(10) if yes, create object *o* from *s* and put *o* in C, and go to (5).
(11) if no, go to (7).

7.4.2 Pros-and-Cons Model

The above mapping algorithm, by assuming a one-one correspondence between functions and structures, is limited in the following aspects:

(1) It does not consider the case where one option agent can fulfil multiple functional requirements simultaneously.
(2) It does not consider the case where one functional requirement can be implemented by a number of option agents.

Attributes and clauses represent the properties of an agent. Some properties describe an agent in terms of static information such as geometry, dimensions, material, weight, etc. Others describe its performance and behaviour, for instance how well it performs if it is used for a purpose under a specific situation. It is the knowledge of the latter type that plays an essential role in conceptual design for selecting appropriate structures (objects). The pros-and-cons model is based on this contextual or subjective type of knowledge, referred to in terms of qualities.

Conceptually, the pros-and-cons knowledge-representation model can be represented in figure 7.5. An entity is implemented by an agent in terms of AGENTS constructs. Entity descriptions include its attributes, pointers to other agent entities, and clauses of relevance. The purpose section represents a set of functions for which this particular entity agent can be used:

PURPOSES$\{$FUNCTION$_i$ | $i = 1, 2, ..., M\}$.

For every function FUNCTION$_i$, there is a set of N_i qualities

QUALITIES$_i\{$QUALITY$_j$ | $j = 1, 2, ..., N_i\}$.

Entity: ENTITY	
Entity Descriptions	
Purposes	
Qualities	
pros	**cons**

Fig. 7.5 conceptual pros-cons model

The set of qualities for a function can be divided into two sub-sets: the pros-set and the cons-set. Qualities in the pros-set can be considered as advantages in favour of a decision to use this entity to implement a function. Qualities in the cons-set can be considered as evidence prejudicial to the decision. In other word, a decision is to balance the positive and negative qualities of an entity.

Qualities can be viewed from a variety of viewpoints. For instance, a decision to use a friction clutch for speed change in a transmission box has a pros-quality that speed change is continuous from the viewpoint of productivity, since continuous speed change reduces unproductive time. However, such a decision may have a cons-quality that heat produced due to friction affects the performance of bearings and causes thermal deformation.

A viewpoint is represented by an advisory agent. The set of advisory agents is called the advisory committee. Different design activities can be monitored by different committees of advisory agents. For example, the advisory committee for conceptual design may be different from that for layout configuration.

The pros-and-cons model allows the positive and negative aspects of a design proposal to be collated and assessed, allowing candidate decisions to be confirmed or refuted.

One of the simplest criteria for justification is based on the net number of positive and negative qualities. A decision with more advantages and/or fewer disadvantages is regarded as superior. This, however, raises some problems. For instance, the importance of individual qualities is neglected. For example, in aeroplane design, reliability and weight are very important factors while cost is less important. On the other hand, in vacuum cleaner design, cost and appearance are important whilst, regrettably, reliability is not always given a high priority.

An alternative, more sophisticated, approach uses a scoring mechanism. Each decision is associated with a total score. The total score of a decision reflects a composite quality including a number of factors. A decision with a higher score is considered to be better.

Advisory Scores

From the point of view of one advisory agent, a decision gains a score, called its Advisory Score (AS). The advisory score AS can be assessed from individual qualities of a decision. For a specific decision, there are m qualities from an advisory viewpoint. Each quality has a score. The sign of the score indicates the type of the quality. That is, if a quality takes a minus score, it is a disadvantage in the cons set; otherwise, it is an advantage in the pros set. The value of the score represents the strength of the quality. The advisory score AS can be calculated as the average of them, i.e.,

$$AS = \frac{\sum\limits_{i=1}^{m} Score}{m}$$

When a decision is made to use a single agent to implement two or more functions at the same time, the advisory score must be computed in a different way. Suppose that there are w functions that can be implemented by a single entity E at the same time.

$$AS = \frac{\sum\limits_{i=1}^{w} \sum\limits_{j=1}^{m_i} Score_j}{\sum\limits_{i=1}^{w} m_i}$$

Committee Score CS

For the committee of the advisory agents, a decision gains a Committee Score (CS). The committee score takes into account the importance of individual advisory agents: every advisory agent $Agent_i$ has an importance weight $Weight_i$. For a committee containing n agents, the CS can be calculated as the average of the total advisory scores considering agent weighting. That is,

$$CS = \frac{\sum\limits_{i=1}^{n} Weight_i \times AS_i}{\sum\limits_{i=1}^{n} Weight_i}$$

Total Score TS

The committee Score CS can be used either to justify (confirm or refute) a decision by comparing with a threshold, or to select an optimal decision by comparing the qualities of alternative decisions. In addition, it can be used to assess the final decision of the conceptual design by adding up the committee scores of all the decisions steps to the Total Score (TS). Suppose that a conceptual design consists of k decision steps, every decision step $Step_i$ having the committee score CS_i. The total score of the design can be calculated as the average of the total committee score:

$$TS = \frac{\sum_{i=1}^{k} CS_i}{k}$$

7.4.3 Principles of Morphological Design

Qualities are in general domain-dependent design knowledge. There are, however, some domain-independent design principles, two of which are discussed here.

Principle of Least-Dependence

Suppose that there are two sets of entities:

functional set F = (F1, F2, ..., Fm)
structural space S = (S1, S2, ..., Sn)

In general, a functional entity can be satisfied by (or "dependent-on") a number of structural entities, and a structural entity can satisfy more than one functional entity. The principle of least-dependence derived from the axiom of functional independence *(Kim 1985)* can be used to resolve an appropriate choice:

A structure with least number of functions is in general preferred.

For example, in the example in figure 7.6, structural entity S1 should be selected for implementing functional entity F1 because fewer functional entities can be feasibly satisfied by S1 than S2 (which also satisfies F1) according to the least-dependence principle. This does not, however, affect the selection of S2 for F2 because there is only one structural entity S2 which satisfies functional entity F2.

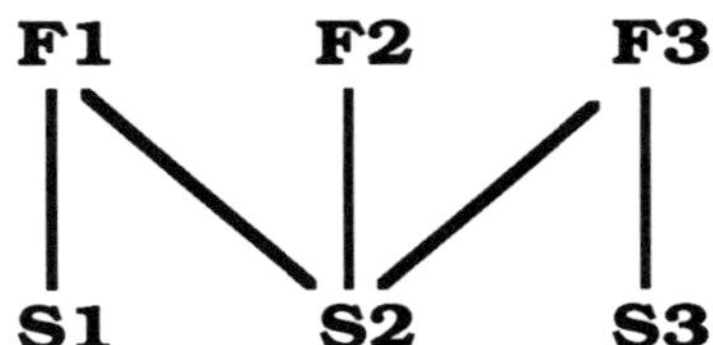

Fig. 7.6 example of least-dependence

Principle of Least Components

However, the validity of the least-dependence principle during conceptual mapping can be affected by other requirements. This is one of the exceptions where a structure can implement two functions at the same time yet both are needed in the design. In this case, the principle of least components can be used to resolve such a conflict:

The number of components in a design should be minimal.

Considering the above example, if structure S2 can serve three functions F1, F2, and F3, and these three functions are all required in a design, only S2 is selected to provide the three functions by the principle of least components.

7.4.4 Morphological Conceptual Design Using Quality Models

Representation of Qualities

A quality can be represented by an AGENTS formalism within an agent of the following general form:

    quality(Function, Advisor, Property, Score).
    quality(Function, Advisor, Property, Score) :- Body.

where

(1) Function – This tells the quality associated with the Function.

(2) Advisor – This gives the information of advisory viewpoint.

(3) Property – This describes the quality.

(4) Score – This specifies strength and type of quality with its sign
    (+ or -) representing advantages or disadvantages respectively;

its value ranges from 0 to 5.

(5) Body – This is used optionally to represent a conditional quality.

Representation of Functions

A function is represented in an agent in the following form:

    function(Name, Description).
    function(Name, Description) :- Body.

where

(1) Name – This specifies the name of the function.

(2) Description – This is a term which describes the function itself. It is not included in this discussion. However, it is worth noting that the description of a function must be consistent with the description of a functional requirement, for instance, of the form motion(rotary, x, cutting).

(3) Body – This is used optionally for representing conditional aspects of the function.

An entity agent can be used to fulfil a number of functions simultaneously. For example, a clutch can be used in a transmission box for speed changes, starting-stopping, and even reversing. In this case, the following representation can be used:

    functions([H | T]).
    functions([H | T]) :- Body.

where

(1) [H | T] – This argument is a list, whose elements represent names of functions which this entity can implement altogether.

(2) Body – This is used optionally to represent conditional situations.

Representation of Advisory Committee and Advice

The advisory committee for a design can be represented by one of the following predicates:

    advisory_committee(Activity, Members).
    advisory_committee(Activity, Members) :- Body.

where

(1) Activity – This describes an activity, conceptual or layout.
(2) Members – This is a list of advisory agents specified by their names.
(3) Body – This is used optionally to represent conditional situations.

Advice can be defined within an advisory agent in a suitable formalism. With the pros-and-cons model, advice by an advisory agent is given to an executive agent on its qualities. However, it is possible for an advisory agent to contribute other types of advice, for example for resolving conflicts. In the next section, advice is represented in terms of constraints within the advisory agents.

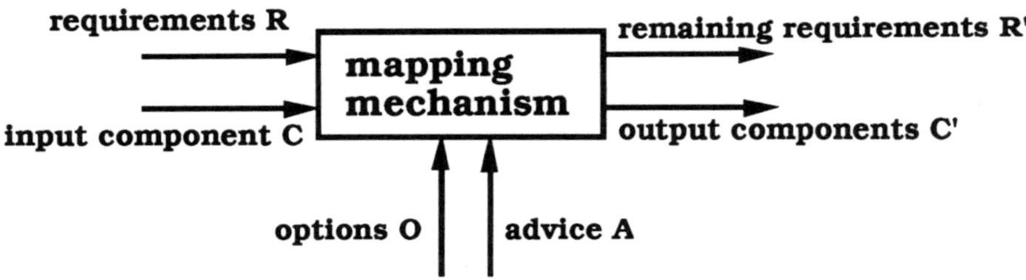

Fig. 7.7 the core mapping mechanism

The Conceptual Design

The mapping system using the pros-and-cons model is based on a central mapping mechanism shown in figure 7.7. It takes two elements R and C as its input and produces two outputs R' and C'.It uses O and A as its resources.

Two termination conditions are possible, either a successful termination or a failure. A complete program   can be found in Appendix C.

(1) A mapping is successful if and only if all the requirements are implemented.
(2) A mapping fails if any one of the requirements cannot be implemented within the available options.

7.5 AGENT-BASED CONSTRAINED CONFIGURATION DESIGN
7.5.1 Constraint Network Representation and Propagation

A product, as a system, comprises a set of objects (components). Layout configuration establishes the relationships between components in the set, especially spatial ones. In terms of graph theory, components can be represented by vertices and relationships by edges connecting vertices. Naturally, the product can then be represented by a graph or a network. Vertices and edges are called graph elements. (See Chapter 3 for more discussions on graphical modelling). Consider, for example, a product which is composed of five components through the stage of conceptual design. As far as the configuration problem is concerned, the five components are represented by five vertices which are variables, as shown in figure 7.8 (a). Every variable vertex has a range of possible values, called labels. The value range is called the label set. The process of assigning a value to a variable element is called labelling. If all the graph elements are labelled, then the graph is said to be labelled.

Two labels are said to be compatible if they are used to label adjacent vertices. Binary compatibilities are useful for establishing relationships (edges) between labelled vertices. For example,

> vertex A is labelled with label a_1,
>> label a_1 is compatible with label b_1,
>> label b_1 is within the label set of vertex B.
> vertex B can feasibly be labelled with b_1.

The above process is shown in figure 7.8 (b). The system continues to look for a compatibility of label b_1 and expands the labelling process in the similar way.

Associated with a compatibility say between a_1 and b_1, there is a set of constraints to be satisfied when an appropriate expansion is to be made. A successful labelling must satisfy all the associated constraints. That is, B can be labelled with b_1 if and only if there is a compatibility between a_1 and b_1 and the associated constraints $C(a_1, b_1)$ are all satisfied. Suppose that:

label b_1 has two compatibilities with c_1 and c_2.

- vertex C is not labelled with c_1 since at least one of the attached constraints is violated.
- vertex C is successfully labelled with c_2 because all the attached constraints are satisfied.

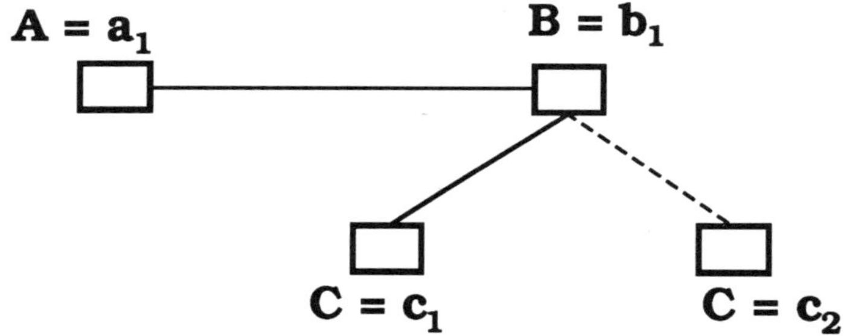

(a) five components as five variable vertices with labels

(b) constraint propagation

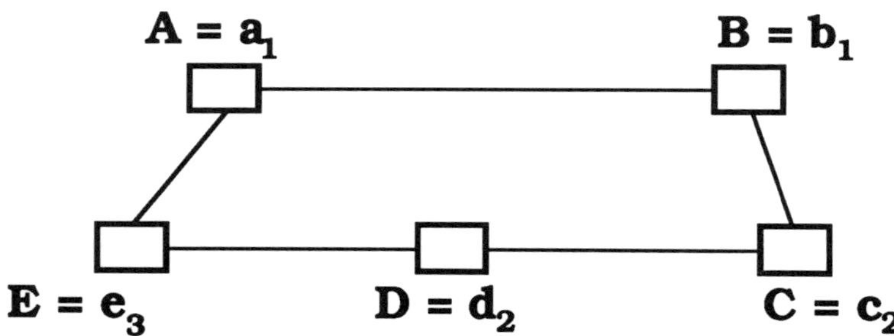

(c) successfully labelled graph

Fig. 7.8 constraint network: representation and propagation

In the similar fashion, labelling continues. The process of labelling under constraints is called constraint propagation.

7.5.2 Agent-Based Constrained Configuration

Representation of Labels

The representation of labels depends on the representation of components resulting from the conceptual design. A label is usually represented by an object since the corresponding component is usually represented by an object. However, agents can also be used for the same purpose.

If a label is represented by an object, it is necessary to indicate the corresponding agent. Therefore, a label may ultimately be represented in the following form:

object(AgentName, ObjectName).

Representation of Graphs

There are a number of ways in which a graph can be represented in Prolog, as discussed by *Bratko (1986)*. A graph can be defined by two sets: its vertex set and edge set. The vertex set represents the set of components which are being configured. The edge set is a list of edges. An edge is a predicate with two labels as its arguments:

edge(Label1, Label2).
edge(object(Agent1, Object1), object(Agent2, Object2)).

where

(1) object(Agent1, Object1) – This represents the first vertex of the edge, that is, the first label Label1.
(2) object(Agent2, Object2) – This represents the other vertex of the edge, that is, the other label Label2.

Representation of Compatibilities

Compatibilities are represented in terms of clauses within label agents of the following form:

```
topology(Next).
topology(Next) :- Body.
?- topology(Next).
```

where

(1) Next – This is a label agent with which the current agent label is compatible.
(2) Body – This is used if a compatibility is conditional.
(3) ?- – This indicates that this compatibility is established by the user.

Representation of Constraints

Constraints are imposed by a committee of advisory agents. The advisory committee of agents is represented by the following predicate:

```
advisory_committee(layout, AgentsList).
```

A constraint is represented within an advisory agent in one of the following forms:

```
constraint(First, Next).
constraint(First, Next) :- Body.
?- constraint(First, Next).
```

where

(1) First, Next – are compatible agent labels.
(2) Body – This is used if the constraint is also conditional.
(3) ?- – This indicates that the constraint is imposed by the user.

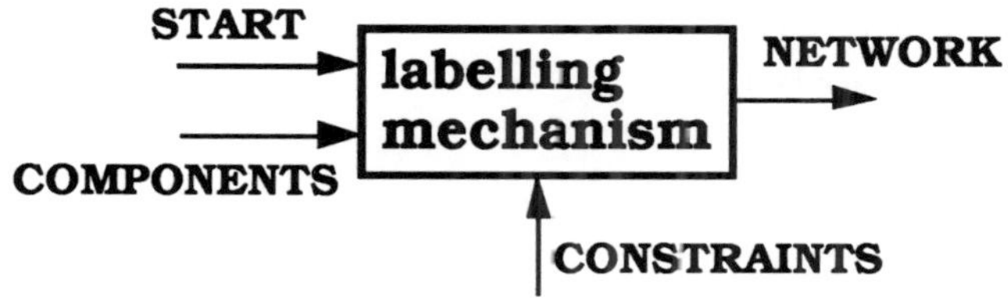

Fig. 7.9 the labelling mechanism

Mechanisms for Configuration

The central mechanism of configuration is that of labelling. The labelling mechanism, shown in figure 7.9, takes two inputs:

(1) The initial label START from which the configuration starts.

(2) The set of components COMPONENTS, each of which must be configured to have a position in the assembly.

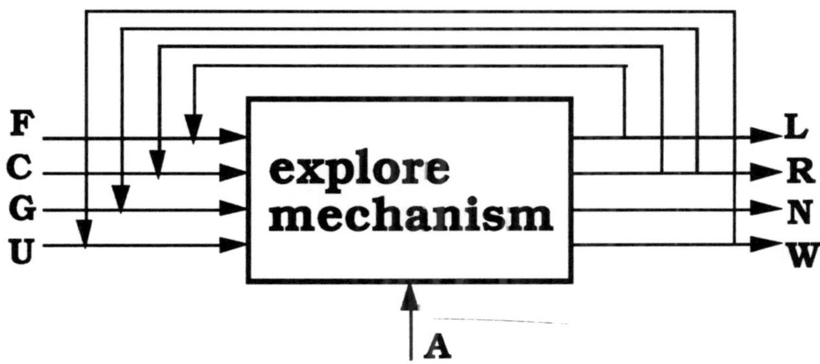

Fig. 7.10 the exploring mechanism

The output from the mechanism is a graph NETWORK. The labelling process is under a set of CONSTRAINTS. The labelling mechanism makes use of another mechanism "explore", as shown in figure 7.10. The exploration mechanism takes the following elements as input:

(1) Current Label F.

(2) Label Set C for the next labelling.

(3) Initial Network G.

(4) Used Labels U.

The output from the exploration mechanism includes:

(1) Next Label L:

(2) Label Set R for next next label:

(3) Resulting Network N:

(4) Used Label W.

Constraints on exploration are represented by the advisory committee of agents A.

Exploration Strategies

The process of constrained exploration is recursive:

(1) The process stops successfully if and only if all the components have been successfully configured. That is, the label set C (R) is empty.

(2) The process fails if there is no label that can be generated. In other words, none of compatible labels of F can lead to a successful labelling.

7.6 SUMMARY

This chapter has used the AGENTS system for developing two typical design systems. The first entails conceptual design using a quality model based on the morphological analysis. The other demonstrates layout configuration using constraint networks. Various aspects of the solution process are discussed:

(1) The discussion ranges over three primary design methods, considering their application to three general classes of design problems at three different stages of the design process. It is assumed that these classifications are distinct. However, not all problems in manufacturing design can be classified in such a way, nor do all problems evolve through these three stages, nor are all problems solved utilising these three particular methods. Both

morphological methods and parametric methods can be used for layout configuration.

(2) The quality (pros-and-cons) model provides a natural means of eliciting knowledge from experts, texts, and/or handbooks. Designers make decisions concerning the selection of a component by appraising some objective function relating to total quality within the context of the product. In some handbooks for manufacturing design, standard or modularised components are assessed in terms of advantages and disadvantages. General scoring of qualities is appropriate in the quality model for morphological conceptual design, in order to resolve conflicts between qualities.

(3) Configuration is considered here in isolation from composition. More generally, however, it is necessary to consider the development of an individual component throughout the process of configuration; it is also necessary to consider the possibility that new components may be added or old components discarded as configuration progresses.

(4) Constraints and compatibilities are two important concepts used in network representations of configuration. Compatibilities, in general, represent domain-specific knowledge which can be observed and captured in advance. Constraints are both domain and problem specific. In this discussion, only binary compatibilities and constraints are included. Compatibilities are represented within the corresponding executive agents, whereas constraints can be separately represented by the advisory committee of agents.

Chapter 8
Conceptual Configuration of Machine Tools

8.1 INTRODUCTION

A systematic study has been conducted by a research team led by Ito at the Tokyo Institute of Technology *(Ito et al 1981-1988)* on conceptual configuration of industrial automation systems, ranging from machine tool structures to entire manufacturing systems. The techniques which were used are rather primitive: digitised description of machine tool structures, symbolised description of machine tool functions, and combinatorial synthesis. As a consequence, the resulting systems have only a limited capability for dealing with domain-specific problems properly, though their ambitious objective is to develop a system able to generate solutions according to requirements from a client. The implication of their research is that AI knowledge-based expert systems may be more advantageous over their previous strategies. This is confirmed by another research team led by Milacic at the University of Belgrade *(Milacic and Pilipovic 1986)*.

This chapter demonstrates a simple example, adapted from a textbook *(Mehta 1984)*, by applying the approach of the agent-based conceptual configuration discussed in the preceding chapter. The primary objective is to evaluate the AGENTS system and its utility in manufacturing design. A secondary objective is to develop an aid that helps with the selection of appropriate components according to functional requirements, the arrangement and relationship of components in space, giving better overall quality, including reduced machining cost, savings in space, material, and weight, and other advantages. An AGENTS program   for demonstration is given in Appendix C. However, substantial further work, especially on eliciting domain knowledge, is required before the system can solve a realistic problem of machine tool design.

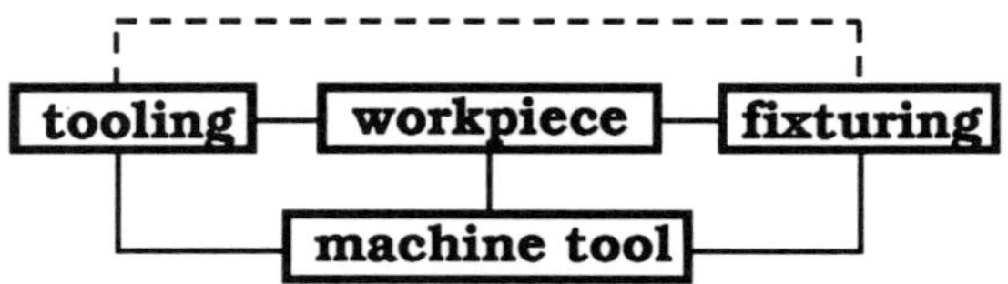

Fig. 8.1 machining environment

8.2 CONCEPTUAL CONFIGURATION

8.2.1 Problem Description

A machine tool works within a machining environment which can be considered to consist of at least four elements: tooling, workpiece, fixturing, and the machine tool itself, as shown in figure 8.1. Therefore, the design of machine tools must also deal with an integrated environment, considering effects from various sources. This chapter, however, considers only two elements: workpieces and machine tools.

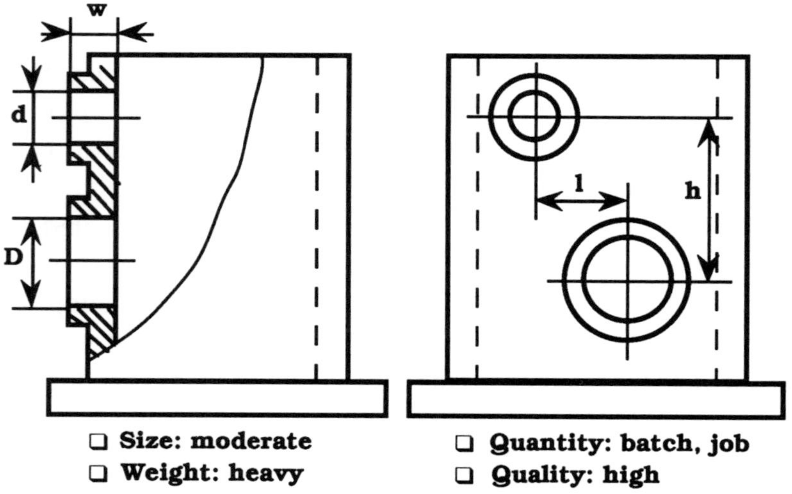

Fig. 8.2 a composite workpiece

8.2.2 Input: Workpiece Information

The input to the system comprises machining features of workpieces, including both geometrical and technological information. This is based on the observation that the design and selection of a machine tool for a specific production system is usually dominated by the machining features extracted from typical workpieces *(Kilmartin and Leonard 1983)*. Figure 8.2 shows an example workpiece, which is of box type.

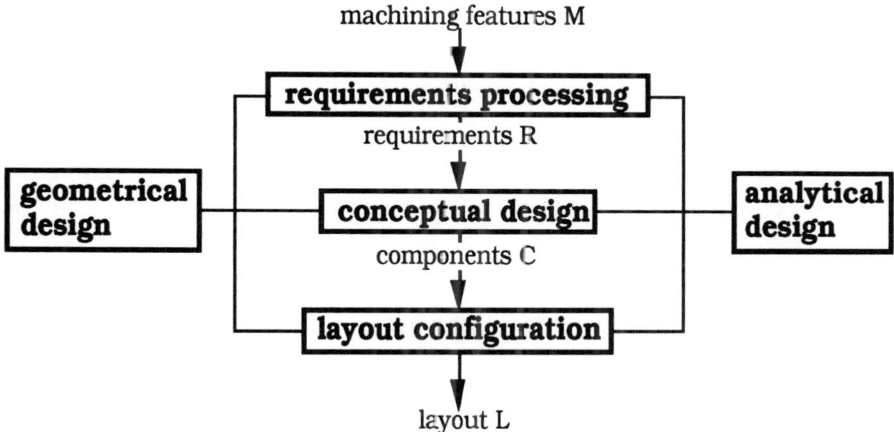

Fig. 8.3 conceptual configuration of machine tools

8.2.3 Tasks

The general process of conceptual configuration can be divided into three steps: requirements processing, conceptual design and layout configuration, as shown in figure 8.3. The requirements processor accepts a set M of the raw descriptions of machining features of workpieces, and produces a set R of functional specifications. At the conceptual step, the requirement set R is transformed into a set C of structural components. The layout configuration step establishes the relationships between elements of the component set C and produces a layout network L.

8.2.4 Characteristics

Conceptual configuration is an early activity during the process of machine tool design although it affects subsequent decisions radically. It is an activity at the conceptual level because it deals with both functional and structural aspects and the mapping between them. It is configurational because it is concerned not just with components but also with their relations. It is ill-structured because available information is incomplete and inexact. It is characterised by diversity because it leads to a large number of alternative solutions. It is heuristic because knowledge is domain dependent and may be company specific, although there are company, national and international standards regarding component design.

Geometrical and analytical aspects of machine tool design are difficult to integrate. Only the most general aspects of component geometry, including shapes, dimensions, and positions, can be considered. Because of the qualitative nature of component representation, it is hardly possible to carry out a design analysis. In this chapter, these two aspects are not included for discussion. It is maintained that they can be integrated with the evolution of design from qualitative to quantitative, from symbolic to numerical.

Conceptualisation and configuration are two different kinds of abstraction. However, knowledge used for conceptualisation and configuration interacts in complicated ways: knowledge for conceptualisation may be useful for configuration and vice versa. A decision from conceptualisation will have effects on subsequent configuration decisions and vice versa. Nevertheless, this chapter makes a distinction that conceptualisation is mainly concerned with the composition of the entity, while configuration deals with the overall structure. As such, they are dealt with independently.

8.2.5 Complexity

The problem of designing a machine tool can be dealt with from a number of different viewpoints from the perspectives shown in figure 8.4. Through the abstraction of projection, complex machine tools can be described in terms of a number of images from different viewpoints with appropriate perspectives. A number of abstraction operations, including projection, can in turn be applied to entity images and their features.

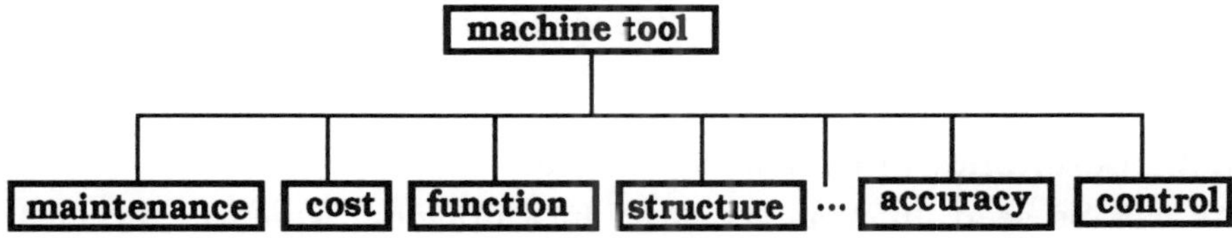

Fig. 8.4 projection viewpoints for machine tool design

The most important yet difficult aspect of using the projection operation is to reveal interacting relationships between those aspects of relevance to the current design. This is also extremely important to enable the organisation of an agent community for solving a design problem in a cooperative environment. Figure 8.5 is the generic network for machine tools including five important components (or modules).

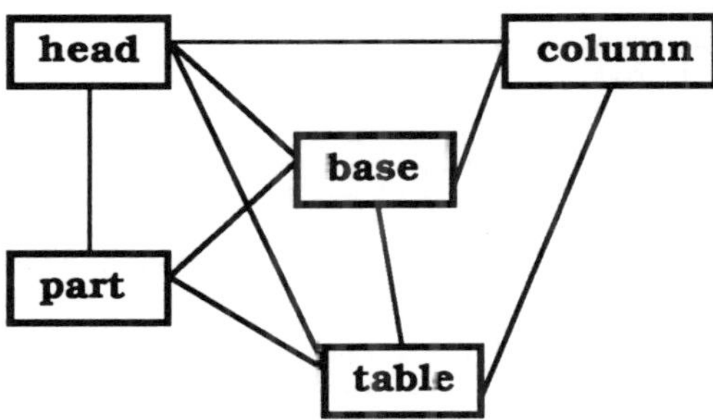

Fig. 8.5 the generic network of machine tools

As far as conceptual configuration of machine tools is concerned, only a sub-set of the viewpoints are actually taken into account, depending on what are thought to be influential. For instance, the issue of accuracy is considered when a machine tool to be designed is used to machine heavy workpieces. In this example, only four aspects are considered. They are: function, cost, structure, and accuracy.

8.3 REQUIREMENT PROCESSING

The objective of requirement processing is to establish the set of functional requirements based on the information about workpieces. Significant difficulties exist if computer systems are to be utilised to assist in this activity, since knowledge at this level is generally extremely heuristic.

8.3.1 Description of Workpiece Features

As can be seen, information given in a drawing, which is the most common medium of communication used by engineers, cannot be directly used in a computer system for conceptual configuration. Instead, it must be represented in certain formalism. Table 8.1 lists machining features relevant to conceptual configuration. In the first column   is a definition of the workpiece agent. In the second column, descriptions are provided.

Table 8.1 machining features of the composite workpiece

workpiece agent	descriptions
agent(workpiece).	This is the workpiece agent.
attribute(size, moderate).	Size is moderate.
attribute(weight, heavy).	Weight is heavy.
attribute(quantity, moderate).	Quantity is moderate.
attribute(quality, high).	Quality is high.
featurehole1, hole, x).	hole1 is a hole machining feature along x axis.
feature(hole2, hole, x).	hole2 is another hole machining feature along x axis.
distance(hole1, hole2, z, h).	Central distance between two holes along z axis is h.
distance(hole1, hole2, y, l).	Central distance between two holes along yaxis is l.
hole(hole1, d1, w).	hole1 has a diameter d1 and depth w.
hole(hole2, d2, w).	hole2 has a diameter d2 and depth w.
base(horizontal).	The workpiece is fixtured on the basis of the bottom face which is horizontal.
endagent.	This is the end of workpiece agent.

8.3.2 Motion Representation

There are six types of motions: three translatory motions along axes x, y and z, and three rotary motions around axes x, y and z. A motion, either translatory or rotary, can be a principal cutting, feed, or an auxiliary motion. A machine can have more than one motion of the same type for different functions. They are summarised in table 8.2.

Table 8.2 motion descriptions

type	orientation	function
translatory rotary	x y z	cutting feed auxiliary

In AGENTS, a motion can be represented by a predicate of the following form:

motion(Type, Axis, Function).
motion(Type, Axis, Function) :- Body.

where

(1) Type – This indicates the type of motion, translatory or rotary.
(2) Axis – This specifies the orientation of the movement, x, y, and z.
(3) Function – This describes the use of motion, cutting, feed, auxiliary.
(4) Body – This is used conditionally.

Some examples of representation of motions in Prolog terms are listed in table 8.3 with their corresponding descriptions.

Table 8.3 motion representations

motion(rotary,x,cutting).	There is a rotary motion around x axis. It is the principal cutting movement.
motion(translatory,y,auxiliary).	There is a translatory auxiliary motion along y axis.
motion(translatory,x,feed).	There is a translatory feed motion along x axis.
motion(translatory,z,_).	There is a translatory motion along z axis. Its type is unknown.

Functions of a structural module in a machine tool are primarily described in terms of the motions that the module can provide, since movements have effects on the formation of geometry of workpieces. Following the formalism for function representation proposed in the preceding chapter, motion functions can then be described as follows:

```
function(Name, motion(Type, Axis, Function)).
function(Name, motion(Type, Axis, Function)) :- Body.
?- function(Name, motion(Type, Axis, Function)).
```

8.3.3 Processing of Functional Requirements

The functional requirements of a machine tool are represented, within the "machine" or "user" agent, by the following clause:

```
requirements([H | T]).
```

Its argument is a list of requirements. Each element in the list is of the following form:

```
motion(Type, Axis, Function).
```

The list of functional requirements is established by executing the procedure for requirement processing. It excludes unnecessary motions from the total set of six motions TM = [x, y, z, rx, ry, rz]. The procedure of requirement processing, possessed by the "machine" agent, are of the typical form:

```
requirement_processing :-
    translate(x),  ;;; test if motion(translatory, x, _) is required.
    translate(y),  ;;; test if motion(translatory, y, _) is required.
    translate(z),  ;;; test if motion(translatory, z, _) is required.
    rotate(x),     ;;; test if motion(rotary, x, _) is required.
    rotate(y),     ;;; test if motion(rotary, y, _) is required.
    rotate(z).     ;;; test if motion(rotary, z, _) is required.
```

Whether a motion is needed in a machine tool is determined by features of workpieces. Considering the example of workpiece in figure 8.1, a rotary motion around the x axis is required as the principal cutting motion since a hole is machined by a boring operation. The boring operation requires a feed (translatory) motion along the x axis. These two rules of design can be represented in the following Prolog clauses:

```
rotate(x) :-
  workpiece ?- feature(Name, hole, x),
  machine ?- retract(requirements(R)),
  machine ?- asserta(requirements([motion(rotary,x,cutting) | R]))).

translate(x) :-
  workpiece ?- feature(Name, hole, x),
  machine ?- retract(requirements(R)),
  machine ?- asserta(requirements([motion(translatory,x,feed) | R]))).
```

Other motions can be tested in a similar way, although test rules may vary. Currently, an incomplete set of rules is used to analyse machining features of workpieces, to decide appropriate machining operations, and finally to select necessary motions as functional requirements. Table 8.4 lists functional requirements for machining the workpiece in figure 8.2.

Table 8.4 functional requirements

---

motion(rotary, x, cutting)
motion(translatory, x, feed)
motion(translatory, z, auxiliary
motion(translatory, y, auxiliary

8.4 KB CONCEPTUAL DESIGN

8.4.1 Structural Entities

A machine tool, as a system, can be divided into sub-systems called machine units at several levels such as machine level, structural module level, component level, and primitive level, from the structural viewpoint. The discussion in this section concentrates on structural modules with no attempt made to discuss specific shapes, dimensions, etc. in detail.

A structural module can be characterised by its key component(s) and/or principal function(s). For example, the key component of the column module is a beam (with one dimension much greater than the other two), and its principal function is to provide a vertical movement. As far as key components are concerned, there are six types of basic structural primitives: beam-like block, beam-like cylinder, box-like block, box-like cylinder, plate-like block, and plate-like cylinder. They are shown in figure 8.6.

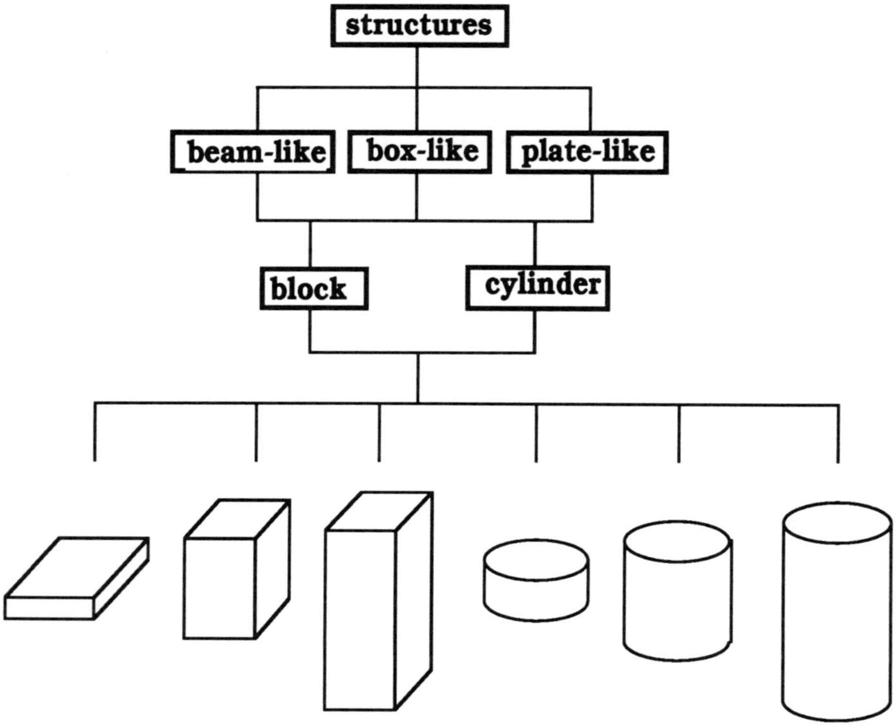

Fig. 8.6 structural primitives

Both functional and structural information of a module are important for conceptual configuration. Table 8.5 lists a number of modules with the key component and the principal function.

Table 8.5 modules with functions and geometry

name	main function	key component	AGENTS representation
column	z translatory	beam-like block	agent((column). motion(t,z,F) :- member(F,[feed,auxiliary]). endagent.
slide table	x, y translatory	plate-like block	agent(table). motion(t,y,F) :- member(F,y,[feed,auxiliary]). endagent.
swivel table	z rotary	plate-like cylinder	agent(w_table). motion(t,y,F) :- member(F,y,[feed,auxiliary]). endagent.
spindle head	x, y, z rotary	box-like block	agent(spindle). motion(r,Axis,cutting) :- member(Axis,[x,y,z]). endagent.
base	stationary	beam-like block	agent(base). endagent.

A structural entity is represented by an agent. An example agent used in conceptual configuration is shown in figure 8.7.

```
agent(column).
      motion(translatory, z, F) :-
            member(F, [feed, auxiliary]).
      topology(base).
      topology(table).
endagent.
```

Fig. 8.7 sample agent in conceptual design

8.4.2 Mapping Results

The morphological conceptual designer discussed in the preceding chapter can be used to determine the components which implement the given requirements. The mapping mechanism considers one functional requirement at a time, sending a message to the group of agents representing candidate entities for functions. If one of the agents can fulfil the function, then it is selected and incorporated into the component set.

Table 8.6 components from conceptual design

```
component(column, column1)
component(head, head1)
component(table, table1)
component(table, table2)
component(base, base1)
```

A simplification, however, is made here, i.e. a one-one correspondence between functions and structures is assumed. This assumption is made mainly due to the lack of domain knowledge currently acquired in the system. Therefore, the conceptual design cannot exploit the quality model. This can be simply superseded by implementing an empty committee of advisory agents:

advisory_committee(conceptual, []).

Table 8.6 presents the result from the conceptual designer with input in table 8.4, options in table 8.5, and an empty advisory committee.

8.5 KB LAYOUT CONFIGURATION

8.5.1 Pairwise Compatibilities

Pairwise compatibilities represent generic connectivities between components and combined functions. For example, the spindle head module is arranged on the side of the column module in order to provide a vertical movement. In AGENTS, a compatibility is represented by a predicate of the following form:

topology(Module1, Module2).

This predicate gives the information that Module1 is adjacent to Module2. The combined function, however, is not explicitly represented for simplicity. It should also be noted that the topological representation is directed from the first (Module1) to the second (Module2). The direction is consistent with that of the flow of forces within a machine tool structure *(Ito et al 1984)*. Table 8.7 lists some of the general connectivity information between modules. They can be defined within corresponding agents.

Table 8.7 some topologies

---

topology(column, base).

topology(head, table).

topology(head, column).

topology(table, table).

topology(table, base).

topology(part, table).

topology(post, head).

topology(part, base).

Labels for vertices are represented by (names of) agents. Therefore, a labeled vertex also represents an agent. Figure 8.7 also shows an example of agent used in configuration.

Table 8.8 constraint agents

constraint 1	constraint 1	constraint 1
IF machine tool is meant for machining heavy workpieces THEN it is not desirable that workpieces should be given a vertical displacement. That is, the column should not be arranged at the workpiece side.	IF heavy workpieces must be machined with high degree of accuracy THEN workpieces should be either stationary or have only one horizontal displacement in order to prevent the effect of workpiece weight on the accuracy. That is, there should be only one table at the workpiece side.	IF heavy and accurate workpieces have a horizontal movement THEN the table carrying the workpieces must be adjacent to the foundation in order to prevent the weight of the moving assembly with workpieces from affecting the machining accuracy. That is, the table should be arranged adjacent to the base.

agent(motion).
  legal(column, Wside) :-
    member(column, Wside),
    !, fail.
  legal(_, _).
endagent
	agent(accuracy).
	  legal(_,Wside) :-
	    member(table1,Wside),
	    member(table2,Wside),
	    !, fail.
	  legal(_,_).
	endagent.
		agent(accuracy, consult).
		  legal(base, Table) :-
		    member(Table,
		      [Table1,Table2]).
		endagent.

8.5.2 Constraints

There are two kinds of constraints: domain dependent and domain independent. Table 8.8 lists three statements of requirements that must be satisfied during the configuration of machine tools.

8.5.3 Network Representation

A graph is defined by two sets: the vertex set and the edge set. The vertex set of the machine tool has been determined during the stage of conceptual design. The edge set is to be established during the configuration.

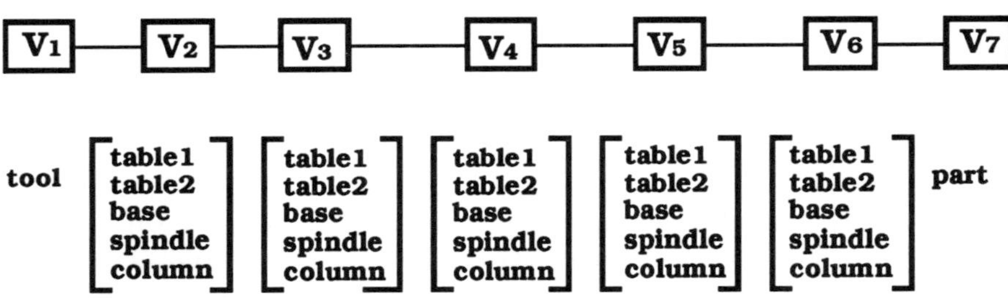

Fig. 8.8 constraint network

The constraint network of this example takes the special form shown in figure 8.9. It consists of seven vertices

$$V = [V1, V2, V3, V4, V5, V6, V7],$$

and six edges

$$E = [E1, E2, E3, E4, E5, E6].$$

It is assumed that (1) only vertices have labels and (2) only edges have constraints. The vertices V1 and V7 have only one element in their respective label sets:

$$Lv1 = [tool], \text{ and } Lv2 = [part].$$

The other five vertices have the same label set:

$$Lv2 = Lv3 = Lv4 = Lv5 = Lv6 = [spindle,table1,table2,base,column].$$

Constraints attached to edges are the same: imposed by a community of advisory agents. However, there is an extra constraint on propagation:

"Any label in the label set can only be used once".

For instance, if V2 is labelled with S, the S cannot be used to label V3 or any others.

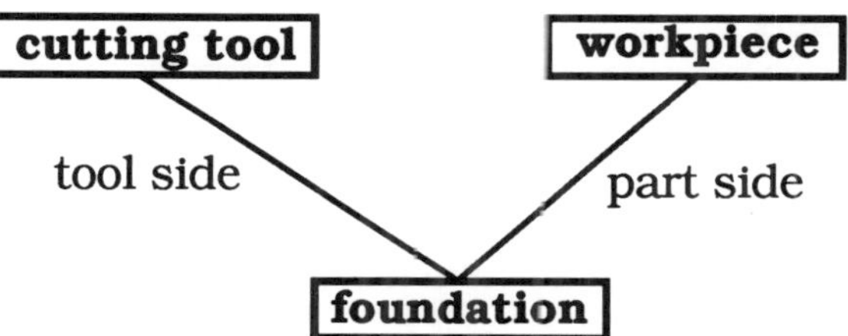

Fig. 8.9 three special vertices to start propagation

8.5.4 Constrained Configuration

Constraint propagation can, in general, start with any component vertex. There are three special vertices: the tool, the workpiece, and the foundation (base), as shown in figure 8.10. In this work, the base module is selected to start the configuration by propagating constraints.

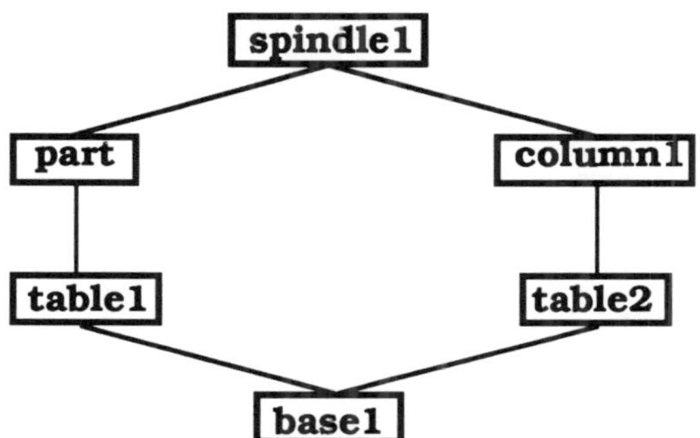

Fig. 8.10 schematic network of the solution

A successful session of configuration will lead to a layout proposal for a machine tool satisfying all the requirements imposed represented as constraints. The quality of the output solution depends on the quality of the knowledge (constraints) used during the configuration. If the constraints used are optimal ones, then the solution is optimal; if they are feasible ones, then the solution is feasible (acceptable but not necessarily optimal); if they represent possibility, then the solution is possible (perhaps not acceptable).

Table 8.9 edges of solution networks

(a) result 1

edge(object(head, head1), object(column, column1))
edge(object(table, table1), object(base, base1))
edge(object(table, table2), object(base, base1))
edge(object(column, column1), object(table, table2))

(b) result 2

edge(object(head, head1), object(column, column1))
edge(object(table, table1), object(base, base1))
edge(object(table, table2), object(column, column1))
edge(object(table, table2), object(base, base1))

Table 8.9 shows two feasible solutions generated by the system, represented by the set of edges representing a graphical network, as shown in figure 8.10. They can be interpreted by two schematic diagrams of possible machine tools, shown in figure 8.11.

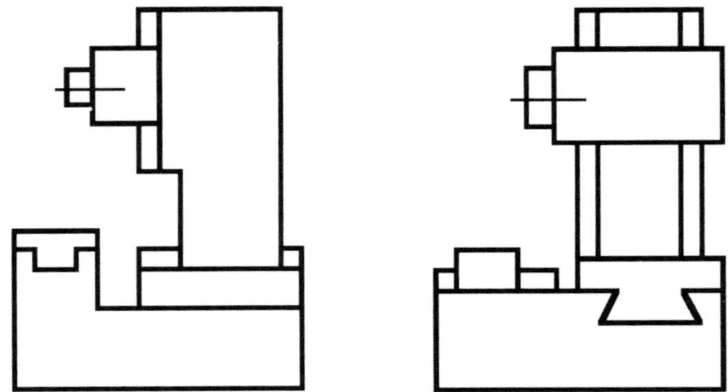

Fig. 8.11 schematic diagram of generated solution

The system reaches its solution from the initial state by applying a sequence of legal operations, which may include activation of other operations belonging to other agents. The first agent (in this case the base agent) starts to expand a vertex and constructs an edge to bridge between itself and the expanded vertex. This action is successful if and only if no constraints associated with this action are violated (Note that a conflict resolution mechanism is not currently implemented). A successfully expanded vertex (which is an object of an agent) repeats the above process until the final conditions are satisfied. Note that the automatic backtracking mechanism of Prolog works for recovery from a temporal failure. For instance, if the current vertex (agent) fails with all of its plans under the current state of the problem, the system will return to the vertex (agent) preceding the current agent to generate another way of expanding and bridging a new vertex. The user can use the explanation facility provided by the AGENTS system to enquire how the system reached its solution.

8.6 SUMMARY

This chapter has illustrated the ideas presented earlier by using as an example the conceptual design of machine tools. Given a set of descriptions of required machining features of a workpiece, the system produces an overall structure of a suitable machine tool. This process is divided into three major stages: requirement processing, conceptual design and layout configuration. It is important to appreciate certain aspects of the problem:

(1) Currently, there is only a limited amount of knowledge in the system, sufficient solely for the purpose of demonstration.
(2) Machine tools are both products of manufacturing systems and components of the systems themselves. Unfortunately, different machine manufacturers have different manufacturing technology, implying that company-specific technological environments must be defined for design systems. Furthermore, different users have different requirements for their specific functional or physical

arrangements in the system. This kind of flexibility has to be balanced by other criteria such as cost constraints.

(3) The machining operations incorporated are turning, milling, drilling and their combination as in machining centres.

(4) Conceptual configuration deals with very coarse-grained aspects of machine tool design. The output from the system is a set of components with their spatial connectivity information specified. However, the system cannot specify the way in which components are assembled. This will result in ambiguity in interpreting the network representation into a schematic representation.

(6) Compared with the approach used by Ito et al(1988), the approach used here is knowledge based, instead of a random combination leading to combinatorial complexity.

Chapter 9
Comparative Appraisal of Alternative Approaches to Cooperative Design

9.1 INTRODUCTION

It is unusual to examine existing design systems near the end of a text, rather than at the outset. In this case, however, the view is taken that the comparative appraisal of alternative approaches, in particular the authors' AGENTS system, is best appreciated once the general and specific aspects of cooperating expert systems are described in detail.

The research on the AGENTS system has been carried out in a breadth-first fashion based on a wide variety of research topics, including distributed problem solving, distributed artificial intelligence, engineering design, database technology, object-oriented programming, logic programming (Prolog), and knowledge-based expert systems. Where systems developed by other researchers are relevant they have been introduced at the appropriate point in the text and these discussions will not be repeated in this chapter. The emphases of this chapter are on the assessment of comparative merits of existing ational frameworks for cooperative work and knowledge - based cooperation in mechanical manufacturing design. Because of the latter restriction, some otherwise interesting research has been excluded, for example studies in civil engineering design such as SMECI by *Haren et al (1985)*.

9.2 COMPUTATIONAL ENVIRONMENTS FOR COOPERATIVE WORK

9.2.1 Expert Systems Shells

Cooperating expert systems can be implemented by commercial expert system shells such as KEE and ART, as has been attempted by *Clarke (1989)* and *Mayer and Lu (1988)*. In general, such environments should provide object-oriented or similar features so that a suitable degree of modularity can be achieved. However, there is a lack of the dedicated features which are required for cooperative work and simple modular components hardly qualify as expert systems.

9.2.2 Sheu's SHEU

Sheu (1988) has combined the deductive Prolog features with some of the object-oriented constructs. Some of these extensions are briefly outlined as follows:

- ***class(a)*** is true if *a* is an object class.
- ***instance(a, b)*** is true if object *a* is an instance of class *b*.
- ***attribute-value(a, b, c)*** is true if object *a* has *c* as a value of its attribute *b*.
- ***execute(c: a(x_1, ..., x_i, y_1, ..., y_j))*** is true if *a* is an operation associated with class *c* and when *a* is executed with the input arguments x_1, ..., x_i and the output arguments y_1, ..., y_j. In this case, it is called a method predicate.
- ***f(x_1, ..., x_i)*** is true if $<x_1, ..., x_i>$ is a member of user-defined relation *f*. In this case, it is called a relational predicate.

There is little evidence, however, that this system can be used as an effective tool to support cooperative activities. It provides only a simple combination of logical constructs with limited object-oriented constructs.

9.2.3 DKOM and ORIENT84/K

Tokoro and Ishikawa (1988) have proposed an approach called DKOM (Distributed Knowledge Object Modelling). In this approach, a knowledge system consists of cooperative knowledge objects, the resulting language being called ORIENT84/K. The language inherits from both Smalltalk-80 and Prolog in terms of its syntax and semantics. There are two kinds of knowledge objects: classes and instances, both of which consist of the following parts:

- **OWN VARIABLES:** This part defines attributes of an object (which may be either a class or an instance).
- **MONITOR PART:** This part fulfils the major functions of control, including access control, prioritised message handling, and statistic gathering.
- **BEHAVIOUR PART:** This part defines methods which respond to messages but is controlled by the information in the monitor part.
- **KNOWLEDGE-BASE PART:** This serves as a local knowledge base of an object. This is supposed to simulate a Prolog programme, containing rules and facts. Rules and facts in the knowledge base of an object are made accessible to methods defined in the behaviour part. A mechanism to call a method of an object from inside the knowledge-base part is also provided.

Although the motivation is to develop a language that can support cooperative work, the resulting system is neither compatible with Smalltalk nor with Prolog and contains very complicated structures and features. The functions of three parts including monitor, behaviour, and knowledge base are rather confusing.

9.2.4 CooperA

CooperA (Cooperating Agents) *(Sommaruga et al 1989)* has been designed and developed as a working prototype software environment supporting the cooperation of heterogeneous distributed semi-autonomous knowledge–based systems. The building blocks include:

(1) CoKernel: This is the cooperation shell, consisting of a collection of initialisation procedures and system facilities that each agent can utilise.

(2) Message-Passing Mechanism: This is a set of functions that handle and control messages. This is the only way that agents communicate and hence cooperate.

(3) CooperA System Agents: This is a set of generic agents built in CooperA which are independent of any application. These generic agents perform operations of common interest of all the application agents.

(4) CooperA Application Agents: These include all those agents that are devoted to one or more application problems. Each agent is a basic ational entity with the capability of fulfilling one or more tasks.

The overall architecture of the CooperA system does not seem complicated with fairly essential components. However, examination of the structure of individual CooperA agents and the way that they communicate reveals the complexity of the CooperA system and its difficulty for users or programmers to master. Although CooperA agents have a well-defined structure, their descriptive attributes are numerous:

> context
> status
> goals
> current message
> has-table
> incoming message queue
> outgoing message queue
> my skills
> unsatisfied goals
> activation method
> plans
> precondition
> yellow pages
> interested in table
> dictionary

9.2.5 CAESAR

The objective of developing the CAESAR (Cooperating Agent Environment for System Activity Representation) system *(D'Angelo et al 1988)* is to provide the end user with a suitable programming environment in which intelligent systems can be developed by implementing a number of independent modules (reasoning agents) whose interaction can be defined by appropriate protocols (mutual knowledge and actions). The system has been structured into several linguistic levels:

- ***HRN (HORN Representation Network):*** a low-level language which can be processed directly by the inference engine.
- ***KCL (Kernel Command Language):*** a low-level resident monitor which allows the system to control agents' interaction.
- ***QDL (Query Definition Language):*** A single-agent user-oriented declarative language.
- ***ACT (Agent Cooperation Task):*** A multi-agent declarative language for representing common knowledge and actions among agents.

Individual agents can be easily defined in terms of the ACT language as follows:

```
+Agent(AgentName).
      <Control Clauses>
      <Protocol Clauses>
+Begin.
      <Knowledge Base>
+End.
```

Each agent is defined by its own knowledge base, inference control, and communication protocol. Every part is structured as a collection of clauses and facts. Therefore, the entire CAESAR system can be viewed as a simple extension of Prolog.

9.2.6 Comparative Discussions

The key features of the systems described above are compared here with the corresponding aspects of the AGENTS system:

(1) The SHEU system is partially object-oriented, containing features such as objects, classes, instances, methods, attributes, etc. However, it is deficient in the relational mechanisms contained in fully object-oriented systems such as message passing, inheritance, etc. In contrast, the AGENTS system is fully object-oriented and provides all the Prolog features.

(2) The SHEU system does not actually provide a modular environment for programming objects. Properties of objects must have pointers, requiring an extra argument in the property declaration. The AGENTS system is highly modular. Agents are defined by two syntax clauses -agent- and -endagent- with anything in between compiled as local properties of the agent.

(3) In the SHEU system, management of objects and object properties does not guarantee to maintain consistencies. Great care must be taken when deleting or adding objects or object properties. In the AGENTS system, individual agents are merely a type of data model. The management of agents is at a higher data level, the agent level, with properties defined and manipulated at the clause level.

(4) Different parts are divided explicitly within an object by certain syntax words in ORIENT84/K. In the AGENTS system, there are also four types of clauses providing different functions but they are not divided explicitly. The order in which they appear within the agent declaration is of no significance.

(5) ORIENT84/K has special syntax and semantics which are not yet familiar to the community. AGENTS inherits those of Prolog enabling experienced users or programmers to master the OO Prolog structure easily.

(6) CooperA has an extremely complicated structure without any compensatory benefits in terms of functionality. The specialised terminology is likely to cause users significant difficulties in designing application programs, leading to doubts concerning

its popular acceptance.

(7) Of all the systems discussed, CAESAR and AGENTS are the two that are most similar, since both are varieties of extended Prolog comprising relatively simple constructs. In CAESAR, separation of clauses of inference control, communication protocol, and knowledge base is unnecessary. It can be speculated that systems of these types are most likely to gain popular acceptance. Of these two, AGENTS is simpler in structure without any apparent functional weaknesses.

(8) Logic Constraint Programming languages are particularly suitable for simultaneous engineering. However, they are seriously limited as ational environments for cooperative work. Some of their features are already incorporated into object-oriented programming systems where demons or active methods can be attached to both methods or attributes. Deductive features are also provided within Prolog systems. All these functions have been combined in AGENTS, by incorporating the logic constraint techniques of Prolog into an object-oriented environment.

9.3 KB COOPERATION IN MANUFACTURING DESIGN
9.3.1 Kim's Theory of Manufacturing Science

A fundamental approach is due to *Kim (1985)* who proposes "mathematical foundations for manufacturing", aiming to develop a methodology generally applicable across all activities from analytical to synthetic design. He discusses various models and techniques suitable for multiple-view modelling of manufacturing systems. A conceptual framework is presented which classifies the study of manufacturing systems in terms of three dimensions: phase (analysis and synthesis), quantification (qualitative and quantitative), and formality (informal and formal). A science of manufacturing, if it is to be truly useful, must address both the analysis and synthesis of manufacturing systems. Analysis deals with the study of the components of a system and their interactions, in order to determine the nature and limits of their behaviour, while synthesis deals with rational principles for designing manufacturing systems.

Informal theories relate to a heterogeneous assortment of facts and ideas, while formal theories provide rigorous, coherent systems. Qualitative models deal with concepts, while quantitative theories incorporate equational relationships. Kim also discusses the possibility of computerising an Axiomatic System.

It should be noted that there are many other theories which are useful in manufacturing design such as the finite element method which are not included in Kim's discussion. The framework, as is acknowledged by Kim, is primarily conceptual with a practical implementation requiring substantial further research in the future. Some of the systems analysed in this review chapter only deal with limited aspects or issues of such a framework.

9.3.2 Brown's AIR-CYL

AIR-CYL *(Brown 1984)* is an expert system developed by Brown to demonstrate a systematic theory of knowledge-based design problem solving in the field of mechanical product design. Brown maintains that expert systems technology is particularly suitable for routine design. The complexity of this type of design problems of routine type can be high due to the large number of components. But it is, both functionally and structurally, relatively fixed and the ambiguity is low. Therefore, a well structured organisation, conceptually isomorphic to the scheme of complexity decomposition, can be established for the problem solver.

A design problem solver consists of a hierarchical collection of design specialists. At the upper levels of the hierarchy are specialists in the more general aspects of the component whilst specialists at the lower levels deal with more specific sub-systems or components. They share a common design database which retains design information. Individual specialists make use of other design agents such as plans, steps, tasks, etc., which may in turn call other specialists.

Control and information is communicated from specialist to specialist by passing messages up and down the hierarchy. There is also local communication between a specialist and its tasks, and between a task and its steps. In other words, communication is strictly hierarchical. Brown believes that the design problem of the routine type is "nearly decomposable" since "the interactions between sub-systems are weak but not negligible". Therefore, communication is predominantly hierarchical.

Two important features of the AIR-CYL system are its plan selection and failure handling. Specialists undertake plan selection in order to influence the flow of control. with different plans producing varying sequences of actions and having different effects on design. This process depends on three types of information: the qualities of the plans themselves, the original requirements from the user, and the situation in which the selection takes place including the state of the current design and the design history.

A failure occurs during a design process carried on between a variety of design agents. A design agent needs to inspect failure reports to identify the type of failure which has occurred and then prescribe an appropriate action for failure recovery. There are in general two ways of recovering from failure. The first is in the plan selection, which requires the abandonment of the current design approach and the selection of a different, and hopefully more successful, plan. The second is revision which is an attempt to use the description of a failure to guide the alteration of a piece of the design in order to alleviate the source of the failure. Usually, the second strategy is used first, and if still unsuccessful then the first plan selection is used.

In AIR-CYL, plan selection and failure recovery are completely the internal concern of individual design agents. They know what plans they possess; they are able to assess the quality of various plans and select the best one; they detect their own failure and are able to determine what went wrong; they attempt to fix it locally within themselves, do so if they can, and report failure only if all attempts fail.

The major contribution of Brown's AIR-CYL system is that it introduces a unique approach to knowledge-based mechanical design: the concept of design agents. Traditionally, expert systems developed for mechanical design have been rule-based. However, the type of agent-based design problem solving offered by the AIR-CYL system is different from those expert systems that use object-oriented techniques. Design agents should be considered as higher-level, self-contained (autonomous), intelligent and interacting entities which act as individual problem solvers. Design agents are able to make use of some AI techniques, such as rule-based, object-oriented, etc., for knowledge representation and inference.

9.3.3 MetaModel

Tomiyama et al (1989) described a system called METAMODEL, implemented at the University of Tokyo. It was one of the first attempts to develop a composite product modelling system for integrating intelligent interactive CAD systems. The research in METAMODEL derived from the work by *Yoshikawa (1981)* who recognised the need, and difficulty, of developing a general purpose product model for different CAD activities. Three basic principles of modelling can be summarised as follows:

(1) A model has a background theory by which the world is modelled.
(2) A background theory consists of a set of concepts and laws concerning phenomena.
(3) A background theory is generic for its instances and the model is complete within the domain where the background theory is valid.

The METAMODEL serves three functions in an intelligent CAD environment:

(1) a central model for integrated product modelling
(2) a mechanism for modelling physical phenomena (functions)
(3) a tool to describe the evolutionary nature of product design.

The underlying rationale of METAMODEL is that various models of a product cannot be independent of each other. In other words, they must share some characteristics, albeit represented differently and perhaps incompatibly. Thus a method has to be provided to propagate changes from one model to another. This area of research plays an important role in cooperative design problem solving since different knowledge sources may well employ different models, methods, and techniques, but they have a significant degree of similarity, since they are different descriptions of the same entities. The central decision is whether to utilise this similarity or neglect it.

In the METAMODEL approach, integration of various models is achieved at the meta level. Three integration principles are summarised as follows:

(1) An object-level model in the METAMODEL system must have its own symbolic representation compatible with the aspect it describes. This symbolic representation is a meta-level model.
(2) A meta-level model is symbolically represented by physical concepts in its background theories.
(3) A METAMODEL system has a system-wide knowledge base about physical phenomena and physical laws to maintain consistency and integrity between models.

Traditionally, product data direct from a CAD system must be processed to a form which can be used for process planning, i.e. geometric information must be converted to manufacturing information, a process which must be carried out explicitly. According to METAMODEL, geometric models and manufacturing models are differentiated at both the object level and meta level. However, the relationships between geometric features and manufacturing features are captured within the METAMODEL so that information transfer can make use of these recorded relationships.

The METAMODEL serves as a computer version of a design handbook for a class of components. Associated properties of components of the same class are stored in the METAMODEL, including functions, structures, and other phenomena. Designers can then retrieve an instance from the component library according to a design specification to generate a solution. Simultaneously, design analysis can be performed since the corresponding background theories are embedded into the METAMODEL and evaluation of models can be automatically generated for analysis.

9.3.4 Chieng's OPTDEX

The OPTDEX System, developed by *Chieng and Hoeltzel (1987)*, offers an interactive hybrid approach to mechanical design. The system incorporates both symbolic rough estimation and numerical evaluation to achieve the optimal design problem formulation and solution. It captures the nature of design problem solving, from qualitative symbols to quantitative numbers, from general representations to specific analytical representations.

This system utilises the concept of design managers. The top-level design manager, a rule-based expert system, is responsible for the overall design. When a design problem passes through the top-level design manager, it breaks down and propagates the global boundary conditions into local boundary conditions. Cell-level design managers are responsible for the design of components which are transmitted from the top-level design manager to constitute the design problem. Their main task is to select a nearly optimal design for a component. An optimisation-level design manager aims to select an appropriate optimisation strategy for a problem and then attempts to implement it. The first half of this task can be rule-based whereas the latter half depends on the chosen optimisation method. Other design managers such as a graphics level design manager and a learning manager are also included.

OPTDEX is a system which has achieved the integration of various design activities including geometrical modelling, analytical optimisation and AI expert systems techniques. Some of the issues are not clear from the published literature, for example, how information concerning these design aspects is represented internally, how the design managers share information and knowledge, how they communicate with each other, how they interact, etc.

9.3.5 Ishii's DAISIE

Ishii et al (1989) have proposed a model of simultaneous engineering design. The resulting system is called DAISIE (Designer's AId for SImultaneous Engineering). The basic methodology behind the DAISIE system recognises that a product is designed in an environment within which a group of experts from disparate disciplines and with different viewpoints look at the problem, collectively proposing a solution to the overall problem. The central concepts used by the system include:

(1) **Design Values:** The concept of design values indicates the different opinions of a candidate design held by corresponding experts who are responsible for the design at different stages of development.

(2) **Design Elements:** Each expert visualises a design candidate as comprising design (building) blocks depending on its design value. The smallest and most fundamental blocks are called design elements.

One of the central ideas of the DAISIE system is the cooperative assessment or evaluation of design proposals in knowledge-based expert design systems. The model makes use of various forms of data, such as compatibilities and their fusion, to generate a consensus opinion of the group. Each expert examines a design proposal, compares it with those existing in its knowledge base to identify a compatibility datum, matches it over a measurable template to obtain a compatibility coefficient, and computes a match index for a design value of the expert. All the experts carries out this process simultaneously and they agree upon a synthesised opinion.

Ishii's model concentrates only on the aspect of proposal assessment within the more general process of design problem solving. There are many other aspects such as proposal generation, and conflict resolution, which are not included in the DAISIE system. This system does not provide or use any comprehensive product model.

9.3.6 Clarke's PROKERN-XPS

The PROKERN-XPS system is an expert system developed by *Clarke (1989)* for the configuration of automation systems. In general, the PROKERN-XPS system operates in two distinct stages:

(1) **Configuration Phase:** During this phase, the configuration is built up by selecting components to fulfil the functional requirements of the automation system.

(2) **Consistency Maintenance:** This phase deals with conflict resolution if there are any inconsistencies.

The PROKERN-XPS system supports multiple knowledge representations by means of a blackboard structure. This is topologically equivalent to a semantic net which allows the relationships between objects and the properties of objects to be efficiently captured. There are three types of network for representing different states of construction:

(1) **Construction Network:** The configuration starts from an initial (static) state represented by the construction network. This contains the static knowledge describing how objects relate to one another and can be combined to create a configuration. It represents all the potential configurations.

(2) **Elaboration Network:** The elaboration network represents the current state of configuration, which is characterised by dynamic knowledge.

(3) **Dependency Network:** The dependency network describes relationships between requirements, configuration decisions and configuration components. Information on dependencies is important for determining future actions, consistency maintenance, and explanation.

The problem solver of PROKERN-XPS includes a set of knowledge sources which have access to the three networks to determine the state of overall configuration. The process of configuration is essentially sequential, implying that there are no explicit interactions between knowledge sources. Although

PROKERN-XPS appears to be an extensive system, it is by no means clear from the report *(Clarke 1989)* how individual knowledge sources represent their knowledge or how they communicate with each other.

9.3.7 Lu's DAI Approach

Mayer and Lu (1988) and *Klein and Lu (1989)* have set out an approach based on Distributed Artificial Intelligence for cooperative design problem solving. They believe that knowledge-based systems can only be realistically applied to engineering design by integrating existing systems for primitive tasks to solve a real and complex problem (which is almost always true in manufacturing design). The first report *(Mayer and Lu 1988)* was not intended to develop new AI tools for cooperative design problem solving, but rather to utilise those already available in the community to demonstrate the ideas. The basic methodology of their approach is based on the following three concepts: knowledge sources, the system, and the world.

Metaphorically, imagine a panel of experts or knowledge sources considering a design problem within a CAD (Computer Aided Design) laboratory. These specialist knowledge sources occupy seats positioned around a large terminal screen (called World) which displays the current state of the design solution. A partition is placed between every pair of seats. By isolating sources in this manner, new sources can be added or old ones removed without disturbing the performance of existing sources.All the participating sources can see the World as a blackboard. They recognise when and where they can contribute by constantly examining the state of the World. The System, which serves a number of central functions such as coordination and recording, identifies experts who wish to contribute, and prepares a list of potential contributors together with their intentions, selecting one at a time to make a contribution. The selected expert is free to introduce its own strategies, which may change the state of the solution. All the participating experts observe the changes and examine alternative actions since they can see the blackboard. They discuss and decide alternatively actions. This cycle of assembling interested experts, selecting one of them, enforcing its strategies, and updating the World continues until the World is well-defined or until no expert cares to comment further. This cycle is analogous to the process in rule-based systems where rules are fired sequentially, but the primary difference is that cooperation between experts is included in this case.

It is worth noting that knowledge sources do not have direct pairwise communication between them, being limited to contact through the medium of the central blackboard World. As a consequence, individual experts cannot control one another since this is the prerogative of the System, not either the knowledge sources nor the blackboard World.

A later paper *(Klein and Lu 1989)* sets out an argument that there is a need for a special-purpose tool for cooperative design problem solving. Based on a protocol analysis, they observed that conflict resolution is central to cooperative design problem solving. Various strategies for resolving conflicts at the level of design agents and within individual agents between tasks have been discussed.

9.4 FUTURE IMPROVEMENTS TO AGENTS

The comments presented here indicate additional desirable features to complement those described in more detail in earlier chapters.

9.4.1 Geometrical Reasoning

Geometric modelling (GM) is one of the CAD activities intended to create, manage, and distribute spatially complete representation of geometric features such as the size, shape, structure, form, and spatial relations of a product. Manufacturing design in mechanical engineering is, in general, geometrically dominated, and much of the knowledge generated and any of the ideas developed are presented and recorded in graphical and geometric forms.

Most domain objects in manufacturing design are physical bodies. Geometric features of domain-dependent and problem-specific objects (agents) may be described by special methods, i.e. a domain agent determines its own geometric features and therefore they can be defined by its methods.

If there is a graphics interface agent, inherited by all the other agents, then graphic modelling methods may be defined by the AGENTS graphics tools. A geometric modelling method may also be defined by an external method.

Computer graphics requires strictly exact details about the geometry of an object, which is rarely satisfied during the early stages of design. Therefore, qualitative geometric reasoning techniques are preferable. For example, dimensions are defined by a range of values instead of exact values. These value ranges are refined to exact values as the design proceeds when, consequently, computer displays are suitable. If this cannot be done, a representative value is chosen only for the purpose of display. It has also been recognised that it is necessary to reason about various features while geometric design is in progress.

9.4.2 Analytical Methods

Design optimisation by numerical analysis is meant to improve a solution according to optimality principles. In most manufacturing design, it is necessary, and becomes possible as design proceeds throughout the conceptual stage, to provide adequate amounts of detailed information required by calculation. It is maintained that, with a set of calculation tools programmed as libraries in the system, agents define and formulate problems in a suitable way, and select an appropriate method from the library for design calculation.

Domain-dependent and problem-specific procedures for design analysis may be defined as methods of agents. Analytical methods can then be called when they are needed, just as are other types of methods. A simple analytical method may be defined in AGENTS. In this case, no interface is needed. An analytical method may alternatively be defined as an external method, which may call domain-independent analytical procedures available in the library. For instance, there can be static and dynamic analysis   programs for the spindle agent as two of its methods. These methods can depend on some design optimisation packages and be interfaced with the AGENTS system. When these methods are necessary and allowable at an appropriate point in the design process, they can simply be called and executed.

9.4.3 Databases

Database Interface: It is, in general, necessary for a design system to interface with a database. For example, the AIR-CYL expert system *(Brown 1984)* uses a database to represent the state of the design problem solving. A special agent may be designed with the ability of managing databases. This database agent can be inherited by all the other agents, so that all the agents in the systems have DBMS (DataBase Management System) functions. The actual design of the database agent may depend on the specific DBMS selected. Suppose the selected database is a relational one and the query language is SQL (Structured Query Language). Then, a selection predicate may be defined as follows:

>   select Clause from Database where Conditions.

or

>   Database ?- Clause :- Conditions.

where

(1) *select, from, where*   – key words for selection.
(2) Clause   – relation and tuples to be selected.
(3) Database   – the name of database.
(4) Conditions   – conditions for selection.

The above selection operation first checks if there is a database defined with the given name. Then, it searches for a relation in the database with the given name and arity. Finally, conditions are evaluated.

    On Top of Databases: As Prolog can sit on the top of a database, the AGENTS system can also sit on the top of a database, with each agent having a virtual database. Limited number of clauses are kept in the agent data structure while the rest of clauses are stored in an external database. In so doing, the capacity of the AGENTS system can be extended. In most systems of database-oriented Prolog, however, only facts (not rules) without free variables are allowed to enter agent databases.

If an agent has a database, a query is processed by first checking if there is an appropriate method defined within the agent data structure. If not, the corresponding database is searched.

Databases for Objects: It is perhaps of more practical value to represent an object of an agent by an external database. The object data structure is primarily introduced to represent components that are designed by agents. The component may be created by the agent as required or already exist as a database provided by manufacturers. For example, the system for transmission design needs information about bearings which are standard products available from the market. There is an expert system for selecting an appropriate type of bearing for a specific use. Just as a human designer selects a bearing from a handbook or catalogue, this expert bearing-selection system makes a decision from a handbook compiled into a database. Databases for objects are simple, containing only two tuples: attribute and value.

9.4.4 User Interface

A good man-machine interface is important for a successful implementation of expert systems. Graphic interfaces have been discussed previously as one important factor. Some others are listed below.

Help Facility: During the execution, the system may need to contact the user for information. There are facilities for explaining "why" a question is asked and "how" a decision is made. In response, the user must provide appropriate answers. If the user is a naive designer who may not know what to answer, the system must be able to provide information about how to answer the question, for example by showing some examples (if the user is not familiar with the way to give an answer), and about what to answer by suggesting some sources to find out answers.

Agent Editor: It would be convenient for the user if the system could display its structure on the screen and to show the whole structure of the agent hierarchy. A more advanced version would show the structure as the system was executing. That is, the contents of the blackboard are displayed on the screen.

Window Management: A window management system would improve the system considerably in terms of user friendliness. A specialist agent may be defined to provide all the window management tools such as making a window, removing a window, shifting between windows, clearing a window, etc. This window management agent would then be made inheritable by all the other agents.

9.4.5 External Methods

The AGENTS system is capable of playing an essential part in cooperating knowledge-based systems as their representatives. Individual knowledge-based systems may be developed directly in this argument language or in some other proof languages. The concept of external methods is proposed to improve the interface between the AGENTS argument language and other proof languages. There are two steps of doing this:

(1) All the proof languages are made available within an environment.

(2) The external methods are defined using this common environment.

(3) A simple interface is provided by the AGENTS system with the environment.

For example, the AGENTS system is currently implemented in POP11 and there is a simple interface between the AGENTS and POP11. An external method may be defined in POP11 within an agent. Within the defined external method, other systems can be easily called, for example if NAG and GINO-F are available in POP11:

```
agent(Agent).

    ...
    FUNCTION(ARGUMENTS)
    external {......}.

    ...
endagent.
```

9.4.6 Efficiency

Distributed processing improves efficiency. Irrespective of the efficiency issue of its implementation, the AGENTS system can improve or be extended to improve efficiency. For instance, in an ordinary Prolog program,   there may be a large number of clauses with the same functor and arity. In an AGENTS program,   the number of clauses with the same predicate may be smaller, and therefore the time spent in futile attempts can be largely reduced.

Another aspect of improving the efficiency would be to implement the AGENTS system in a parallel processing environment. Consider   an example of the following clause:

```
Head :- Receiver1 ?- Body1,
        Receiver2 ?- Body2.
```

Currently, *Body1* and *Body2* are proved sequentially. After *Body1* is proved within *Receiver1*, the proof of *Body2* is considered by *Receiver2*. In Concurrent AGENTS, if implemented, *Body1* and *Body2* would be processed concurrently by different processors within respective receivers *Receiver1* and *Receiver2*.

9.4.7 Other Improvements

In the above sub-sections, a number of immediate improvements are discussed. There are more fundamental issues remaining to be solved in the long term. Some of them are listed in the following:

- How to formulate, describe, decompose, and allocate problems and synthesise results among a group of agents.
- How to enable agents to communicate and interact: what and

when to communicate and in what way, etc.

- How to ensure that agents act coherently in making decisions or taking actions, accommodating the global effects of local decisions and avoiding unforeseen harmful interactions.

- How to enable agents to represent and reason about their actions, plans, goals, and knowledge of other agents in order to coordinate them.

- How to reason about the state of the cooperative process, initiation, continuation, and termination.

- How to recognise and reconcile disparate viewpoints and conflicting intentions or decisions among a collection of agents.

9.4.8 Fully/Truly Cooperating Expert Systems

In the examples used to illustrate the system here the agents are so small that they hardly qualify for expert systems although they have the standard KB structure. Recall that one of the objectives of the AGENTS system is to cooperate truly and fully separate and heterogeneous expert systems, preferably developed by different parties. This objective can only be achieved with realistic requirements imposed on the development of individual systems. Future research should be directed towards specific issues raised in applications.

9.5 SUMMARY

Current practices in knowledge-based cooperative mechanical design are limited by a number of factors. The most significant of these is the essential compromise between the provision of a dedicated environment with a relatively simple structure which is capable of supporting a variety of strategies for cooperative problem solving. Most existing environments for cooperative problem solving are, however, both very complicated in their structure and difficult for design engineers to master their most useful features. The AGENTS system is designed to overcome these difficulties although much remains to be investigated.

Chapter 10
Conclusions

Over recent years, a large number of expert systems have appeared for mechanical manufacturing and design. At about the time when the research reported in this text started, research on cooperating knowledge-based systems began receiving increasing and formal attention from the community of manufacturing design research. To date, a small number of systems can incorporate cooperation as one of their essential ingredients only to a limited extent.

This text starts from the identification of the distributed nature in manufacturing design, and proceeds to its management. An intelligent integrated framework has been proposed to simulate, support, and emulate how a group of expert designers cooperate and coordinate to solve a problem collectively. The practicability of the framework has been proved by the successful implementation of the AGENTS system which inherits much from existing expert systems in manufacturing design. In addition, it allows distributed representation of knowledge and enables cooperation by sharing of knowledge through inheritance and message passing. The utility of the AGENTS system in manufacturing design has been demonstrated for morphological modular design and constrained layout configuration. The simple illustrative example of machining centre design can be generalised to other kinds of automation systems. As a result of the efforts in implementing the AGENTS system and using it for manufacturing design, much has been learnt. Although the implementation is not complete, a better understanding has been gained about the strengths of, and problems with, the approach. There are many aspects of cooperative design problem solving and cooperating knowledge-based systems worth further investigation. These have been identified and discussed throughout the text.

An approach to cooperating knowledge-based systems for manufacturing design has been proposed. Much work remains to be done in this area. It is crucial to understand more about the distributed nature of domain complexity, about how it can be managed, about how to incorporate the essential ingredient of cooperation, and about how best to develop systems to achieve these. However, the AGENTS system, and the theory behind it, have revealed some aspects about the essence of integrating disparate knowledge sources during stages of conceptual design and layout configuration of automation systems in manufacturing design, while discovering many interesting and difficult issues for future research.

References

Akao, Y (Ed) (1990) *Quality Function Deployment – Integrating customer requirements into product design,* Productivity Press, Cambridge, Massachusetts, USA

Allen, R H, Boarnet, M G, Culbert, C J, and Savely, R T (1987) "Using Hybrid Expert System Approaches for Engineering Applications" *Engineering with Computers,* Vol 2, No 2, 95-110

Barbuceanu, M (1984) "Object-centred representation and reasoning: An application to computer-aided design" *SIGART Newsletter,* January, 33-39

Bloor, M S, de Pennington, A, Harris, S B, Holdsworth, D, McKay, A, and Shaw, N K (1988) "Towards Integrated Design and Manufacturing Systems" In: *Proceedings of IERE International Conference Factory 2000,* Cambridge, UK, 21-29

Bond, A, and Gasser, L (1988a) "An Analysis of Problems and Research in Distributed Artificial Intelligence" In: *Readings in Distributed Artificial Intelligence,* Bond, A, and Gasser, L (eds), Morgan Kaufmann Publishers, California, USA, 3-35

Bond, A, and Gasser, L (eds) (1988) *Readings in Distributed Artificial Intelligence,* Morgan Kaufmann Publishers, California, USA

Bond, A (1989) "The Cooperation of Experts in Engineering Design" In: *Distributed Artificial Intelligence,* Volume 2, Gasser, L, and

Huhns, M N (eds), Pitman Publishing Ltd, London, UK, 463-483

Borrow, D G (1985) "If Prolog is the answer, what is the question? Or what it takes to support AI programming paradigms" *IEEE Transactions on Software Engineering*, Vol 11, No 11, 1401-1408

Boyer, C (1989) *A History of Mathematics*, John Wiley, New York, USA

Brandon, J A,(1992) *Managing Change in Manufacturing Systems*, Productivity Publishing, Olney, Buckinghamshire, England

Brandon, J A and Huang, G Q (1993) "Use of an AGENT based system for Concurrent Mechanical Design", In:*Concurrent Engineering: Automation, Tools and Techniques*, Kusiak, A (Editor), John Wiley, New York, USA

Bratko, I (1986) *Prolog programming for artificial intelligence*, Addison-Wesley Publishing Company, London, UK

Brown, D C (1984) *Expert systems for design problem solving using design refinement with plan selection and redesign*, PhD Thesis, Ohio State University, USA

Byte (1986) *Byte Magazine*, Special issue on Object Oriented languages, Vol 11, No 8, August

Cammarata, C, McArthur, D and Steeb, R (1983) "Strategies cooperation in distributed problem solving" In: *Proceedings of 8th International Joint Conference on Artificial Intelligence*, 767-770

Chang, E (1987) "Participant Systems for Cooperative Work" In: *Distributed Artificial Intelligence*, Huhns, M N (ed), Morgan Kaufmann Publishers, California, USA, 311-339

Chieng, W H,and Hoeltzel, D A (1987) "Interactive hybrid (symbolic-numeric) system approach to near optimal design of mechanical components" *Engineering with Computers*, Vol 2,

No 2, 111-123

Chu, Q Q, Yao, G Q, Li, D L, and Zhao, R J (1987) "A CAD/CAM System for Typical Component in Mechanical Structure Design" *Annals of the CIRP*, Vol 36, No 1, 61-64

Clarke, B R (1989) "PROKERN-XPS: A mixed architecture expert system for industrial automation system configuration" In: *Proceedings of the 4th International Conference on Applications of Artificial Intelligence in Engineering*, Cambridge, UK, 49-76

Clocksin, W F, and Mellish, C S (1981) *Programming in Prolog*, Springer-Verlag, London

D'Anelo, A, Mirolo, C, and Pagello, E (1988) "A multi-agent planner for reasoning with incomplete knowledge in a distributed environment", In: *Proceedings of European Conference on Artificial Intelligence*, Munich August 1-5, 528-533

Davis, R, and Smith, R G (1983) "Negotiation as a metaphor for distributed problem solving" *Artificial Intelligence*, Vol 20, 63-109

de Kleer J (1986) "An Assumption-based TMS" *Artificial Intelligence*, Vol 28, 127-162

Decker, K S (1987) "Distributed problem-solving techniques: a survey" *IEEE Transactions on Systems, Man, and Cybernetics*, Vol 17, No 5, 729-740

Deitz, D (1989) "The Power of Parametrics" *ASME Mechanical Engineering*, Vol 111, No 1, January, 58-64

Doyle, J (1979) "A Truth Maintenance System" *Artificial Intelligence*, Vol 12, 231-272

Drake, S, and Fela, S (1989) "A Foundation for Features" *ASME Mechanical Engineering*, Vol 111, No 1, January, 66-73

Drucker, P F (1989) *The New Realities*, Heinemann, Oxford, UK

Duda, R, Gaschnig, J, and Hart, P (1979) "Model Design in PROSPECTOR Consultation System for Mineral Exploration" In: *Expert Systems in the Micro-Electronic Age*, Michie, D (ed), Edinburgh University Press, Edinburgh, UK, 153-167

Eder, W E (1980) *Principles of Engineering Design*, written by Hubka, V, translated by Eder, W E, Butterworth Scientific, England

Englemore, R, and Morgan, T (eds) (1988) *Blackboard Systems*, Addison-Wesley Publishing, Wokingham, England

Fagan, M J (1987) "Expert systems applied to mechanical engineering design – experience with bearing selection and application programme" *Computer-Aided Design*, Vol 19, No 7, 361-367

Feng, X A, Li, X L, and Mattias, E (1987) "Intelligent CAD of Shafts, Sleeves, and Cylindrical Gears by Micro Personal Computer" *Annals of the CIRP*, Vol 36, No 1, 69-72

Fikes, R, and Kehler, T (1985) "The Role of Frame-based Representation in Reasoning" *Communications of the ACM*, Vol 28, No 9, September, 904-920

Forsyth, R (1986) "Architecture of Expert Systems" In: *Expert Systems: Principles and Case Studies*, Forsyth, R (ed), Chapman and Hall Publishing, London, UK, 9-17

Fukuda, T, Takeda, S, and Hayashi, M (1986) "Distributed expert systems for production control" In: *Proceedings of IIE Conference*, 222-228

Gasser, L, and Huhns, M N (eds) (1989) *Distributed Artificial Intelligence*, Morgan Kaufmann Publishers, California, USA

Goldberg, A, and Robson, D (1983) *SMALLTALK-80: The language and its implementation*, Addison-Wesley Publishing, Massachusetts, USA

Gordon, W J J (1961) *Synectics*, Harper & Row Inc. N.Y.

Guide (1984) *A guide to design for production*, Institution of Production Engineers, London, UK

Harary, F (1969) *Graph Theory*, Addison-Wesley Publishing, Wokingham, England

Harada, M, Kunii, T L, and Saito, M (1978) "RGT: the recursive graph theory as a theoretical basis of a system design tool DESIGN-TOOL – with application to medical information system design" In: *Proceedings of International Symposium on Medical Information Systems*, Osaka, Japan, 503-507

Haren, P, Neven, B, Giacometti, J P, Montalban, M, and Corby, C (1985) "SMECI: Cooperating expert systems for civil engineering design" *SIGART Newsletter*, Vol 92, 1985, 67-69

Hatvancy, J (1976) "The distribution of functions in manufacturing systems" In: *Proceedings of International IFIP/IFAC Conference on Programming Languages for Machine Tools, PROLAMAT 76*, McPherson, D (ed), VOL 1, North-Holland Publishing, Amsterdam, The Netherlands, 23-28

Hatvancy, J, and Stone, B J (1987) "User friendly CAD" *Annals of the CIRP*, Vol 36, No 2, 451-453

Hayes-Roth, F, Waterman, D, and Lenat, D B (eds) (1983) *Building Expert Systems*, Addison-Wesley Publishing, Massachusetts, USA

Heragu, S S, and Kusiak A (1987) "Analysis of Expert Systems in Manufacturing Design" *IEEE Transactions on Systems, Man, and Cybernetics*, Vol 17, No 6, 898-912

Hinde, C J, Bray, A D, Herbert, P J, Launders, V A, and Round, D (1989) "A truth maintenance approach to process planning" In: *Proceedings of the 4th International Conference on Applications of Artificial Intelligence in Engineering*, Cambridge, UK, 171-188

Huang, G Q. and Brandon, J A (1988a) "Topological representations for machine tool structures" In: *Proceedings of 27th International Machine Tool Design and Research Conference*, Davies, B J (Ed), Manchester, UK, 173-178

Huang, G Q, and Brandon, J A (1988b) "An investigation into machine representation models for machine tool design" In: *Proceedings of International Conference on Factory 2000*, Institution of Electronic and Radio Engineers, Cambridge, UK, 331-336

Huang, G Q, and Brandon, J A (1988c) "Machine tool analysis and synthesis based on the graphical machine representation model" In: *Advances in Manufacturing Technology*, Vol 3, Worthington, B (ed), Kogan Page, London, UK, 100-104

Huang, G Q, and Brandon, J A (1991) "Specification and Management of the Knowledge Base for Design of Machine Tools and Their Integration into Manufacturing Facilities" In: *Artificial Intelligence in Design*, Pham, D T (ed), Springer-Verlag Publishing, London, UK, 169-190

Huang, G Q, and Brandon, J A (1992a) "AGENTS: A Prolog system for cooperating expert systems" *International Journal of Knowledge Based Systems*, Vol 5, No 2, 1992, 125-136

Huang, G Q, and Brandon, J A (1992b) "Cooperating expert systems for CAD/CAM", In: *Proceedings of 29th International MATADOR Conference*, Manchester, UK, 69-74

Huang, G Q, and Brandon, J A (1993) "AGENTS for cooperating expert systems in concurrent engineering design", *International Journal of Artificial Intelligence for Engineering Design, Analysis and Manufacturing*, Vol 7, No 2

Huber, G P (1984) "Issues in the design of group decision support systems" *Management Information Systems* (MIS Quarterly), September, 195-204

Huhns, M N (Ed) (1987) *Distributed Artificial Intelligence*, Morgan Kaufmann Publishers, California, USA, 1987

Ishii, K, Geol, A, and Adler, R E (1989) "A model of simultaneous engineering design" In: *Proceedings of the 4th International Conference on Applications of Artificial Intelligence in Engineering*, Cambridge, England, 484-501

Ito, Y, and Saito, Y (1981) "Computer-aided draughting system 'ALODS' for machine tool structures" In: *Proceedings of 22nd International Machine Tool Design and Research Conference*, Manchester, England, 69-76

Ito, Y, and Shinno, H (1982) "Structural description and similarity evaluation of the structural configuration in machine tools" *International Journal of Machine Tool Design and Research*, Vol 22, No 2, 97-110

Ito, Y, (1984a) "Description of machine tools and its applications - CAD system for machine tools structures" *Bulletin of Japan Society of Precision Engineering*, Vol 18, No 2, 178-185

Ito, Y, and Shinno, H (1984b) "Generating method for structural configuration of machine tools" *Transactions of JSME*, Vol 50, No 449, 788-793

Ito, Y, Shinno, H, and Saito, H (1988) "A proposal for CAD/CAM Interface with expert systems" *International Journal of Robotics and Computer Integrated Manufacturing*, Vol 4, No 3/4, 491-497

Kawagoe, K, and Managaki, M (1984) "Parametric Object Model: A Geometric Data Model for Computer Aided Engineering" *NEC Research and Development*, No 72, January, 23-32

Kilmartin, B R, and Leonard, R (1983) "Selecting advanced machine tools by a systems approach based on key machined components" *Proceedings of Institution of Mechanical Engineers*, Vol 197B, 261-269

Kim, S (1985) *Mathematical foundations of Manufacturing science*, PhD Thesis, MIT, USA

Klein, M, and Lu, S C Y (1989) "Conflict resolution in cooperative design" *International Journal of Artificial Intelligence in Engineering*, Vol 4, No 4, 170-180

Kornfeld, W A, and Hewitt, C E (1981) "The scientific community metaphor" *IEEE Transactions on Systems, Man, and Cybernetics*, Vol 11, No 1, 24-33

Kusiak, A (1990) *Intelligent Manufacturing Systems*, Prentice-Hall Publishing, New York, USA

Laventhol, J (1987) *Programming in POP-11*, Blackwell Scientific Publications, Oxford, UK

Martin, R R (1991) "Geometric reasoning for computer aided design", In: Pham, D T (Ed), *Artificial Intelligence in Design*, Springer-Verlag, London, UK, 47-59

Mayer, A K, and Lu, S C Y (1988) "An AI-based approach for the integration of multiple sources of knowledge to aid engineering design" *Transactions of the ASME, Journal of Mechanisms, Transmissions, and Automation in Design*, September, Vol 110, 316-323

McCorduck, P (1979) *Machines Who Think*, W H Freeman, New York, USA

Mehta, N K (1984) *Machine Tool Design*, TaTa McGraw-Hill Publishing Company Limited, New Delhi, India

Milacic, V, and Pilipovic, M (1986) "Conceptual design based on the linguistic approach and the automata theory" *Annals of the CIRP*, Vol 35, No 1, 1986, 103-106

Minker, J (ed) (1989) *Foundations of deductive databases and logic programing*, Morgan Kaufmann Publishers, California, USA

Nilsson, N J (1980) *Principles of Artificial Intelligence*, Tioga

Publishing Company, California, USA

O'Hare, G M P (1990) "Distributed Artificial Intelligence: An Invaluable Technique for the Development of Intelligent Manufacturing Systems" *Annals of the CIRP*, Vol 39, No 1, 485-488

Pahl, G, and Beitz, W (1977) *Engineering Design*, edited by Wallace, K N, The Design Council/Springer-Verlag, London, UK

Parunak, H V D (1988) "Manufacturing Experience with the Contract-Net" In: *Distributed Artificial Intelligence*, Volume 1, Huhns, M N (ed), Pitman Publishing Ltd, California, USA, 285-310

Peckham, J, and Maryanski, F (1988) "Semantic Data Models" *ACM Computing Survey*, Vol 20, No 3, 153-189

Rendeiro, J O (1985) "How the Japanese Came to Dominate the Machine Tool Business", *Long Range Planning*, 18(3), 62-67.

Roseman, M A, Gero, J S, Hutchinson, P J, and Oxman, R (1986) "Expert systems applications in computer aided design" *Computer Aided Design*, Vol 18, No 10, 546-551

Sharp, J A (1987) *An Introduction to Distributed and Parallel Processing*, Blackwell Scientific, Oxford, England

Shaw, N K, Bloor, M S, de Pennington, A (1989) "Product Data Models" *International Journal of Research in Engineering Design*, Vol 1, No 1, 43-50

Sheu, P C, and Kashyap, R L (1988) "Programming robot systems with knowledge" *International Journal of Robotics and Computer-Integrated Manufacturing*, Vol 4, No 3/4, 359-367

Shortliffe, E H (1976) *Computer-Based Medical Consultation: MYCIN*, American Elsevier Publishing, New York, USA

Smith, G W, and Wang, M (1988b) "Modelling CIM systems Part 2: the generic functions" *Computer-Integrated Manufacturing*

Systems, Vol 1, No 3, 169-178

Smithers T (1989) "AI-based design versus geometry-based design or why design cannot be supported by geometry alone" *Computer-Aided Design*, Vol 21, No 3, 141-150

Smithers, T, Conkie, A, Doheny, J, Logan, B, and Millington, K (1989) "Design as intelligent behaviour: An AI in design research programme" In: *Proceedings of the 4th International Conference on the Applications of Artificial Intelligence in Engineering*, Cambridge, UK, 294-334

Sommaruga, L, Avours, N M, and Liedekerke, V (1989) "An Environment for Experimentation with Interactive Cooperating KB Systems" In: *Proceedings of 9th Annual Technical Conference on Research and Development in Expert Systems VI*, Shadbolt, N (ed), British Computer Society Specialist Group on Expert Systems, Cambridge University Press, Cambridge, UK, 104-115

Stallman, R, and Sussman, G (1977) "Forward Reasoning and Dependency Directed Backtracking in a System for Computer-Aided Circuit Analysis" *Artificial Intelligence*, Vol 9, 135-196

Steeb, R, and Johnston, S C (1984) "A computer-based interactive system for group decision making" *IEEE Transactions on Systems, Man, and Cybernetics*, Vol 11, No 8, 544-552

Stefik, M and Bobrow, D (1986) "Object-Oriented Programming: Themes and Variations" *The AI Magazine*, Vol 6, No 4, Winter, 40-61

Stefik, M, (1981) "Planning and Meta-Planning (MOLGEN: Part 2)" *Artificial Intelligence*, Vol 16, 141-170

Stefik, M (1981) "Planning with Constraints (MOLGEN: Part 1)" *Artificial Intelligence*, Vol 16, 111-140

Su, S Y W (1986) "Modelling integrated manufacturing data with SAM" *Computer*, Vol 19, No 1, 34-49

Subrahmanyam, P A (1985) "The 'software engineering' of expert systems: Is Prolog appropriate?" *IEEE Transactions on Software Engineering*, Vol 11, No 11, 1391-1400

Suzuki, H, Kimura, F, and Sata, T (1986) "Variational product design by constraint propagation and satisfaction in product modelling" *Annals of the CIRP*, Vol 35, No 1, 75-78

Svensson, N L (1974) *Introduction to Engineering Design*, Pitman Publishing, London, UK

Tokoro, M, and Ishikawa, Y (1988) "An Object-Oriented Approach to Knowledge System" In: *Readings in Distributed Artificial Intelligence*, Bond, A, and Gasser, L (eds), Morgan Kaufmann Publishers, California, USA, 425-433

Tomiyama, T, Kiriyama, T, Takeda, H, Xue, D, and Yoshikawa, H (1989) "Metamodel: A key to intelligent CAD systems" *International Journal of Research in Engineering Design*, Vol 1, No 1, 19-34

Tong, C (1987) "Toward an engineering science of knowledge based design" *International Journal of Artificial Intelligence in Engineering*, Vol 2, No 3, 133-166

Topping, B H V, and Kumar, B (1989) 'Knowledge representation and processing for structural engineering design codes" *International Journal, Engineering Application of Artificial Intelligence*, Vol 2, No 3, 1989, 214-227

Ulrich, K, and Seering, W (1988) "Computation and Conceptual Design" *International Journal of Robotics and Computer-Integrated Manufacturing*, Vol 4, No 3/4, 1-7

Ulrich, K, and Seering, W (1989) "Synthesis of schematic descriptions of mechanical design" *Research in Engineering Design*, Vol 1, No 1, 3-18

Waterlow, J G, and Monniot, J P (1986) "A study of the state of the art in computer aided production management", *SERC ACME*

Directorate, Swindon, UK

Wright, P K, Englert, P J, and Hayes, C C (1991) "Application of artificial intelligence to part setup and workholding in automated manufacturing", In: Pham, D T (Ed), *Artificial Intelligence in Design*, Springer-Verlag, London, UK, 295-314

Yoshikawa, H (1981) "General Design Theory and a CAD system" In: *Proceedings of the IFIP Working Conference on Man-Machine Communication in CAD/CAM*, Sata, T, and Warman, E A (eds), North-Holland, Amsterdam, The Netherlands, 35-58

Zarefar, H, Lawley, T J, and Etesami, F (1986) "PAGES: A Parallel Axis Gear Drive Expert System " In: *Proceedings of ASME Conference on Computers in Engineering*, 1986, Vol 1, Gupta, G (ed), USA, 145-149

Appendix A
Backus Naur Form Specification for AGENTS

A.1 INTRODUCTION

The current try-run version of the AGENTS system closely resembles the syntax and semantics of Prolog, while it is fully object-oriented. As has been discussed in chapter 6, an AGENTS program can be viewed in two different ways which will be specified in BNF terms.

A.2 BNF SPECIFICATION FOR STANDARD PROLOG

An AGENTS program can be considered just as an ordinary Prolog program, consisting of clauses. A simplified BNF specification for standard Prolog is given in the following:

```
<program >      ::= {<predicate>}*
<predicate>     ::= {<clause>}*
<clause>        ::= <rule> | <fact> | <query>
<rule>          ::= <head> :- <body> .
<fact>          ::= <head> .
<query>         ::= ?- <body> .
<head>          ::= <structure>
<body>          ::= <structure> {, <structure>}*
<term>          ::= <constant> | <variable> | <structure>
<constant>      ::= <number> | <atom>
<functor>       ::= <atom> | .
<structure>     ::= <functor> <aguments>
<arguments>     ::= (<term>{, <term>}*) | <* empty *>
<atom>          ::= <small>{<character>}* | <string>
<variable>      ::= <capital>{<character>}* | _{<character>}*
<string>        ::= '{<character>}*'
<character>     ::= <letter> | <number> | <symbol>
<number>        ::= <integer> | <real>
```

```
<integer>      ::= <digit>{<digit>}*
<real>         ::= <integer>.<integer>
<letter>       ::= <small> | <capital>
<small>        ::= a | b | ... x | y | z
<capital>      ::= A | B | ... X | Y | Z
<digit>        ::= 0 | 1 | ... 8 | 9
<symbol>       ::= + | - | * | / | \ | ^ | < | > | ~ | : | . | ? | @ | £ | $ | &
```

A.3 BNF SPECIFICATION FOR AGENTS

There is another way of understanding an AGENTS program,   being considered to consist of a number of agents, each of which can be considered as an ordinary Prolog program:

```
<program >        ::= {<agent>}*
<agent>           ::= agent ( <atom> ) .
                         {<predicate>}*
                  endagent .
```

A.4 ASSUMPTIONS

<term>	Names of language constructs are surrounded by "<" and ">".
{X}*	Represents zero or more repetitions of X.
[X]	Means that X is optional.
X | Y	Indicates that X and Y are alternatives and either X or Y must be used.

Appendix B
The POPLOG Core Prolog System

B.1 INTRODUCTION

The experimental implementation of the prototype AGENTS system is in POP11 within the POPLOG AI programming environment on VAX under VMS. POPLOG provides a core Prolog system which is used for implementing the POPLOG Prolog. Details about the POPLOG system and the POPLOG Core Prolog system can be found in the POPLOG system documentation. In this appendix, only a number of features of the POPLOG Core Prolog system are mentioned.

B.2 Prolog Data Types

The data type PROLOGTERM represents general complex terms in Prolog. A Prolog term is characterised by a functor which must be an atom or a string, and optionally a set of arguments. The number of arguments of a Prolog term is called arity.

The data type PROLOGVAR is used to represent variables in Prolog terms. In the POPLOG Core Prolog system, the structure of a PROLOGVAR has a single field containing the value of the variable.

B.3 THE PROLOG MEMORY AREA

B.3.1 The Continuation Stack

The POPLOG Core Prolog is based on the model of "continuation passing". The principle of continuation passing is that: if a procedure successfully completes its computation, it invokes its continuation describing whatever computation needs to be performed; if it fails, it returns.

Consider a rule clause whose body has N terms. Each term corresponds to a continuation procedure or c-procedure which is pushed onto a continuation stack or c-stack together with its arguments if there are any. Then, the c-stack is invoked for execution. The last continuation procedure which corresponds to the first term in the rule body is applied. If this succeeds, the second c-procedure is activated for execution.

This process takes place until the Nth c-procedure has also succeeded, which implies the final success of the goal clause. This is demonstrated in figure B.1. A failure of any c-procedure will cause a return from the continuation stack.

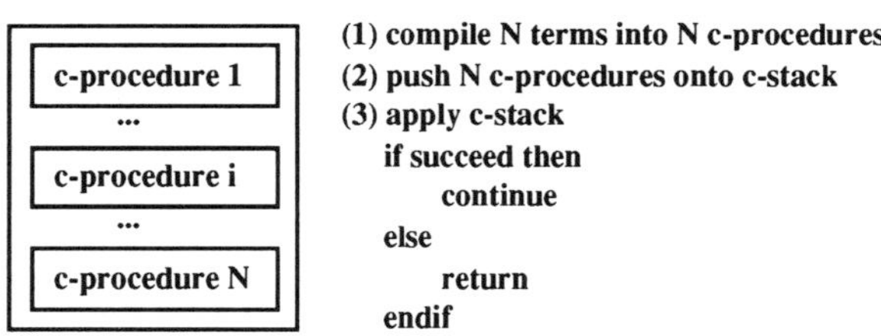

Figure B.1 Prolog procedures using continuation stack

B.3.2 The Trail

The trail stack records all assignments made to Prolog variables. Whenever one of the Prolog variables is assigned to, it is pushed onto the trail stack. At each choice point, a pointer to the current top of the trail is saved. Whenever the Prolog system backtracks through a choice point, the old trail pointer is restored and any variables pushed onto the trail after that point are reset to be undefined.

Another effect of backtracking is to free Prolog variables which are no longer accessible to active procedures. A free-list of Prolog variables is maintained for this purpose: once backtracking has returned past the point at which a Prolog variable was first created, it is put back onto the free-list for later re-use.

B.4 PATTERN UNIFICATION/MATCHING

Conceptually, terms can be represented as tree. Figure B.2 represents an arithmetic expression of the generic form:

$$+(A, *(B, C)).$$

Consider now the first case in figure B.2 (a), where *val1, val2, val3* are constants, and *x* and *y* are two Prolog variables. It is intuitively obvious that, in order to make two trees equal, *x* must be instantiated by *val3*, and *y* by *val1*. Note that *val2* in two trees are accidentally coherent (same value at the same place).

Now, consider the second case in figure B.2 (b), where constant *val2* is replaced by variables *p* and *q* in two trees. This time, *x* and *y* are instantiated as above. Nevertheless, *p* and *q* must be instantiated to the same variable object. Therefore, they must be linked, so that instantiation of the one will lead to instantiation of the other.

Finally, suppose that *q* is also a structure -*(r, s)* where *r* is a Prolog variable and *s* is a constant. In this case, *x* and *y* are instantiated as in the first case, but *p* is instantiated in a special way: by a structure with a constant s and a variable r.

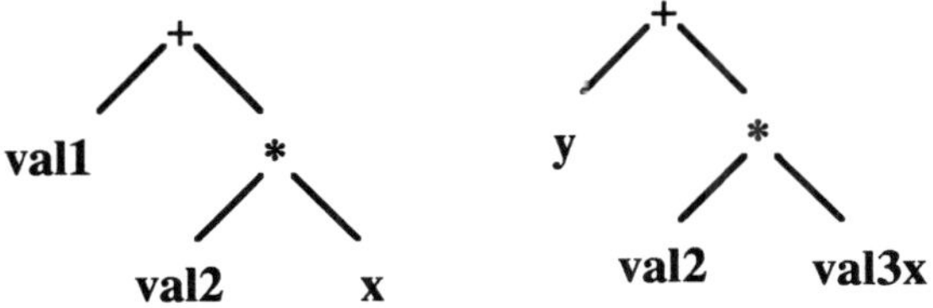

B.2 (a) constant-variable instantiation

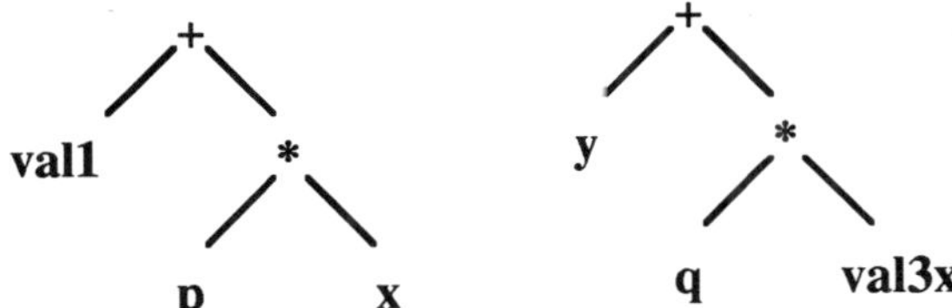

B.2 (b)variable-variable instantiation

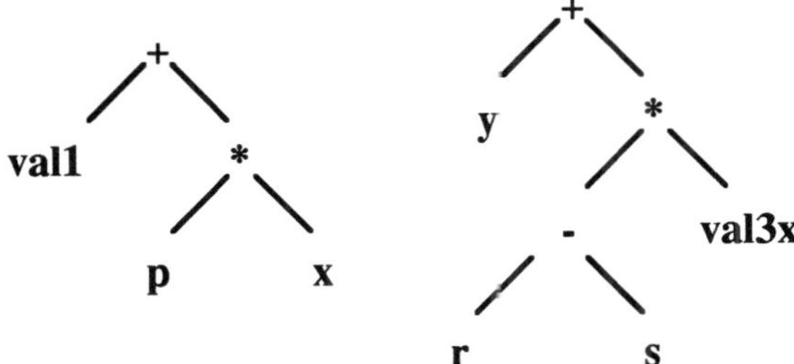

B.2 (c) structure-variable instantiation

Figure B.1 variable instantiation

The general rules to decide whether two terms, S and T, match are as follows:

(1) If S and T are constants then S and T match only if they are the same object.

(2) If S is a variable and T is anything, then they match, and S is instantiated to T. Conversely, if T is a variable then T is instantiated to S.

(3) If S and T are structures then they match only if

    (3.1) S and T have the same principal functor, and

    (3.2) all their corresponding components match.

The resulting instantiation is determined by the matching of components.

B.5 BACKTRACKING AND CUT

The backtracking is marked by a choice point which indicates the state of a Prolog procedure by three variables: the continuation stack pointer, the trail pointer and a pointer to the next free Prolog variable. Before a goal is evaluated, a backtracking point is marked or saved. Any failure of proving this goal will restore the most recent backtracking point. Cut, in general, is used to stop backtracking for alternatives, regardless of all the backtracking points marked for a goal.

Appendix C
List of Sample AGENTS Programs

C.1   AGENTS PROGRAM   FOR PRELIMINARY DESIGN: REQUIRE.AGT

```
agent(require).

preliminary(F) :-
        require ?- asserta(requirements([])),
        translate(x),
        translate(y),
        translate(z),
        rotate(x),
        rotate(y),
        rotate(z),
        !,
        require ?- requirements(F).

rotate(x) :-
        part ?- feature(Name, hole, x),
        retract(requirements(R)),
        asserta(requirements([motion(rotary, x, cutting)|R])).

translate(x) :-
        part ?- feature(Name, hole, x),
        retract(requirements(R)),
        asserta(requirements([motion(translatory, x, feed)|R])).

rotate(y).

rotate(z).
```

```
translate(y) :-
        part ?- distance(Feature1, Feature2, y, L),
        retract(requirements(R)),
        asserta(requirements([motion(translatory, y, auxiliary)|R])).

translate(z) :-
        part ?- distance(Feature1, Feature2, z, H),
        retract(requirements(R)),
        asserta(requirements([motion(translatory, z, auxiliary)|R])).

endagent.
```

C.2   AGENTS PROGRAM   FOR CONCEPTUAL DESIGN: CONCEPT.AGT

```
agent(concept).

conceptual(C) :-
        user ?- functional_requirements(R),
        executive ?- option_table(E),!,
        mapping(R, E, C).

mapping(R, E, C) :-
        mapping(R, E, [], C).
mapping([], E, C, C).
mapping([H1|T1], E, [O|C], CC) :-
        implement(H1, E, O),
        mapping(T1, E, C, CC).

implement(_, [], _) :- fail.
implement(R, [H|T], O) :-
        implement1(R, H, O).
implement(R, [H|T], O) :-
        implement(R, T, O).

implement1(R, H, object(H, O)) :-
        H ?- R,
        create_object(H, O).

endagent.
```

C.3 AGENTS PROGRAM   FOR CONFIGURATION DESIGN: LAYOUT.AGT

```
agent(layout).

layout(Graph) :-
        user ?- structural_components(C),
        start_from(First, Rest, C), !,
        configuration(First, Rest, Graph).

start_from(H, T, [H|T]).

configuration(Start, Components, Graph) :-
        explore(Start, Components, Rest, [], Graph1),
        explore(Start, Rest, [], Graph1, Graph).

explore(Start, [], [], G, G).
explore(object(space, _), C, C, G, G).
explore(object(A, O), C, R, G, Graph) :-
        A ?- topology(Next),
        getone(Next, C, One, L),
        committee_layout(A, Next),
        explore(object(Next, One), L, R,
                [edge(object(A, O), object(Next, One))|G], Graph).

getone(_, [], _, _) :- !, fail.
getone(X, [object(X, Y)|T], Y, T) :- !.
getone(X, [H|T], Y, [H|Z]) :- !,
        getone(X, T, Y, Z).

committee_layout(E, S) :-
        advisory ?- committee(layout, C),
        committee_advisory(C, E, S).

committee_advisory([], _, _).
committee_advisory([H|T], E, S) :-
        expert_layout(H, E, S),
        committee_advisory(T, E, S).

expert_layout(E, A1, A2) :-
        E ?- constraint(A1, A2).

endagent.
```

C.4 COMMUNITY AGENTS IN MACHINE DESIGN: MACHINE.AGT

```
agent(user, consult).

functional_requirements(
        [support,
         motion(rotary, x, cutting),
         motion(translatory, x, feed),
         motion(translatory, z, auxiliary),
         motion(translatory, y, auxiliary)]).
structural_components(
        [object(base, base1),
         object(table, table1),
         object(column, column1),
         object(space, space1),
         object(table, table2),
         object(head, head1)]).

endagent.

agent(executive).
        option_table([base, table, column, head]).
endagent.

agent(advisory).
        committee(conceptual, []).
        committee(layout, [accuracy, motion]).
endagent.

agent(part).
        size(moderate).
        weight(heavy).
        quantity(moderate).
        quality(high).
        feature(hole1, hole, x).
        feature(hole2, hole, x).
        distance(hole1, hole2, z, h).
```

```
        distance(hole1, hole2, y, 1).
        hole(hole1, d1, w).
        hole(hole2, d2, w).
        base(xy).
endagent.

agent(space).
;;; empty agent
endagent.

agent(head).
        function(motion(rotary, XYZ, ANY)).
        topology(space).
endagent.

agent(table).
        function(motion(translatory, x, Any)).
        function(motion(translatory, y, Any)).
        topology(space).
        topology(column).
        topology(table).
endagent.

agent(column).
        function(motion(translatory, z, Any)).
        topology(head).
endagent.

agent(base).
        function(support).
        topology(space).
        topology(column).
        topology(table).
        topology(head).
endagent.
```

```
agent(accuracy).

        constraint(base, table).
        constraint(base, space).
        constraint(base, All_Others) :- !, fail.
        constraint(_, _).

endagent.

agent(motion).

        constraint(column, table) :- !, fail.
        constraint(column, space) :- !, fail.
        constraint(base, space) :- !, fail.
        constraint(_, _).

endagent.
```

Appendix D
Reference Guide to AGENTS

D.1 INTRODUCTION
The AGENTS system is supposed to provide all the standard Prolog predicates,though some of them,especially those for debugging,are currently omitted for the simplicity of implementation without losing the generality, and some of them have not yet been tested due to the time limitation. In this appendix, some of the AGENTS built-in predicates are selectively and briefly discussed in terms of their major functions.

D.2 EXTENDED PREDICATES
In this section, some important predicates built in AGENTS are listed. Most of them are extended object-oriented constructs:

    agent(AgentName)
    agent(AgentName, Flag)
    attribute(Attribute)
    attributes(Attributes)
    class(Agent)
    default(Attribute, Value)
    endagent
    inherit(Agent1, Agent2)
    instance(Agent, Object)
    instantiate(Attribute, Value)
    invite(Agent)
    inviting(Agent)
    listing(Predicate)
    object(Agent, Object)
    plist(Plist)
    sub(Agent)
    subs(Subs)
    super(Agent)

        supers(Supers)
        value(Attribute, Value)
        Head := Body
        Head =: Body
        Head -: Body
        Agent ?- Message

D.3 ALPHABETICAL DIRECTORY OF BUILT-IN PREDICATES

In this section, most of the agents built-in predicates are listed in the alphabetical order with concise explanations of their principal functions.

abolish(Functor, Arity)
cancels a predicate with the functor of Functor and arity of Arity.

abort
stops execution and returns you to the ?- prompt.

agent(AgentName)
starts to define an agent with the name of AgentName which must be an atom. This must be used together with -endagent- as a pair of syntax words for agent definition.

agent(AgentName, Flag)
starts to define an agent with the name of AgentName. The consult flag Flag is used to control whether or not existing agents should be overridden or expanded.

arg(N, Struct, Args)
gives you the value of the Nth Arg(ument) in Struct(ure).

asserta(Clause)
adds whatever is in Clause before other matching database predicates.

assertz(Clause)
adds whatever is in Clause after other matching database predicates.

atom(Atom)
checks whether Atom is a Prolog atom.

atomic(X)
X is a constant (an atom or a number).

attribute(Attributes)
defines an Attributes, which must be atoms, with undefined default values.

attribute(Attribute, DefaultValue)
defines an attribute named Attribute with the default value of DefaultValue.

bagof(Term, Goal, List)
collects all instances of Term which occur in Goal, and puts them into a list List.

call(Goal)
calls the goal in Goal.

class(Agent)
succeeds if and only if Agent is an agent. It is different from agent(Agent) which will create an agent if there is no such an agent.

clause(Head, Body)
Find a clause with the head of Head and body Body.

compile(File)
reads the contents of the file named File into the memory.

default(Attribute, Value)
returns the default value for the named attribute.

endagent
ends a definition of an agent started by -agent-.

fail
This is used to force backtracking with failure.

findall(Term, Goal, List)
constructs a list List consisting of all of the objects Term such that the goal G is satisfied, regardless of possibly different solutions for variables in P that are not shared with X.

functor(Term, Functor, N)
analyses a term into its functor and number of arguments.

get0(Char)
read any character Char from the current input stream.

halt
stops AGENTS and returns to the operating system.

inherit(Agent1, Agent2)
Agent1 inherits from Agent2. It fails if there is no such inheritance.

instance(Agent, Object)
Object is an instance of Agent. It is different from object(Agent, Object) which will create an instance of the agent if the agent does not have such an instance.

instantiate(Attribute, Value)
Attribute is instantiated with Value. It is different from value(Attribute, Value) which checks that Attribute has Value.

integer(Int)
succeeds if Int is instantiated to an integer.

invite(Agent)
change the current agent contributor to Agent.

inviting(Agent)
returns the current agent.

length(List, Length)
counts the number of items in List.

listing
lists out to the current output stream all the clauses of an agent.

listing(Pred, Arity)
lists out all the clauses for the predicate name you specify in Pred of an agent.

name(Atom, List)
List is the list of ASCII codes of the characters that compose the constant Atom.

nl or nl(N)
causes AGENTS to go onto a new line or N new lines in the current output stream.

novar(Var)
succeeds if Var is not an uninstantiated variable.

not(Goal)
succeeds if the attempt to call whatever is in the variable Goal fails and fails if the goal is successfully called.

object(Agent, Object)
Object is an instance of Agent; or create Object as an instance of Agent.

op(Precedence, Associativity, Op)
defines an atom as an operator with the given Precedence and associativity.

plist(Plist)
returns the precedence list of an agent.

predicate(Functor, Arity)
Check if there is any predicate defined within the current agent with the functor of

Functor and arity of Arity.

read(String)
read the next term Term. This must be terminated by a period followed by a carriage return.

repeat
When backtracking, each time AGENTS returns to repeat it stops backtracking and treats all the goals which come after repeat as though it were coming to them for the very first time.

retract(Clause)
deletes the first clause which matches the given pattern Clause.

see(Stream)
set a new current input stream.

seeing(Stream)
returns the current input stream.

seen
This predicate returns AGENTS to the default current input stream - the keyboard.

setof(Term, Goal, List)
collects all instances of Term which occur in Goal and puts them into a list List whose elements are sorted and duplicate elements are deleted.

sub(Agent)
define Agent to be a sub-class agent.

subs(Subs)
returns all the sub-class agents.

super(Agent)
define Agent to be a super agent.

supers(Supers)
returns all the super-class agents.

tab(N)
Output the number of spaces specified by N to the current output stream.

tell(Stream)
changes the current output stream to whatever is specified in the variable Stream.

telling(Stream)

tells you what the current output stream is.

told
makes the current output stream revert to the default - screen.

true
This goal always succeeds.

value(Attribute, Value)
Attribute has Value without instantiation.

var(Var)
succeeds if Var is a variable.

write(Term)
prints something Term on the screen.

!
This is the cut mechanism for controlling backtracking.

X is Y
This is the forced unification between X and Y with evaluation.

X = Y
pattern matching between X and Y without evaluation.

X == Y
pattern matching with evaluation.

X =.. Y
converts a list into a predicate and vice versa.

X > Y or X < Y
Comparison between X and Y.

Head -: Body
define a failure handling demon.

Head := Body
define an after demon.

Head =: Body
define a before demon.

Agent ?- Message
Message is targeted to Agent.

Appendix E
Glossary of Selected Terminology

E.1 GENERAL TERMINOLOGY

manufacturing design: A term used loosely to refer to activities related to both design of products and manufacturing systems.

CAD-G (computer-aided geometrical design), CAD-A (computer-aided analytical design): Applications of computer-aided design in different design activities using different techniques.

AI systems, knowledge-based systems, intelligent systems, ES systems: These concepts are used in this work interchangeably.

information, data, knowledge, expertise: There are differences between these types of concepts. However, they are not strongly distinguished in this work, for the simple reason that their representations share strong similarities.

CAD-I (intelligent computer-aided design): A computer system that supports design using AI techniques.

cooperative KB systems, cooperating ES systems, distributed ES systems, distributed KB systems: These concepts are used in this work interchangeably.

Distributed AI: This is a branch of AI, dealing with situations where multiple agents are involved.

E.2 TERMINOLOGY USED IN COMPLEXITY MANAGEMENT

abstractions: In chapter 3, a variety of abstraction operations are discussed for managing complexity.

divide and conquer: This is a strategy for complexity management, by breaking a problem down into a number of pieces, and then capturing descriptions about and relationships between them.

pros-and-cons model, quality model: They refer to a product model for representing features with associated qualities.

specialisation: The process of modifying a generic thing for a specific use by defining a more specific sub-class object.

configuration: It is mainly the act of determining the layout.

composition: The composition of an artifact includes components that constitute the artifact.

layout: The layout usually refers to relationships between components of an artifact.

scheme: A scheme includes both composition and layout.

embryonic composition: Classes of components that constitute an artifact.

embryonic layout: Relationships between classes of components.

embryo: An embryo includes both embryonic composition and embryonic layout.

generic network: A schematic network is a graph whose vertices represent classes of components and edges general relationships between them.

schematic network: An elaboration network is a graph whose vertices represent instances of components and edges relationships between them.

E.3 TERMINOLOGY USED IN OO PROLOG

AGENTS: This is the name for an Object-oriented Prolog developed in this research.

agents: The agent metaphor is used in this work for several meanings. For instance, an agent is a processing unit; an agent represents a data structure for classes; an agent represents a knowledge source.

objects: The concept of objects is loosely used to refer to every thing. However, a special meaning is to represent instances of classes.

messages: This is how the objects communicate with each other and how the outside world communicates with an object. In terms of AGENTS's Prolog semantics, a message is a query asked by the user or another agent to a specific agent in the system about whether a particular instance of a relation holds. In terms of procedural semantics, a query can be considered as a procedural call.

methods: This is what an object uses to respond to a particular message. Methods are predicates defined by clauses.

attributes: Attributes are variables which describe an object. Objects may have the same attributes but with different values.

inheritances: Inheritances are transferring of similarity in properties from one or more object to another. Inheritance relationships are determined by object abstraction networks.

Constraints or demons: Constraint methods are executed during the object compiling time.

meta agent: This is the special agent used in the AGENTS system.

user agent: This is the special agent used in AGENTS to refer to the human user.

invitation: In AGENTS, an agent can be invited to perform a sequence of activities.

soliloquy: In AGENTS, an agent that is processing a message can be used to process another message.

E.4 TERMINOLOGY USED IN COOPERATIVE DESIGN

cooperative design, concurrent design, simultaneous design, integrated design, total design, distributed design: These concepts are used interchangeably in this work.

blackboard, bulletin board: They are metaphors of a common place for keeping information.

Index